Chinese Broadband and Next-Generation Internet

www.royalcollins.com

Chinese Broadband and Next-Generation Internet

Editors:
Cao Jiguang, Zhao Feng, and Ma Junfeng

Books Beyond Boundaries

ROYAL COLLINS

Chinese Broadband and Next-Generation Internet

Cao Jiguang, Zhao Feng, and Ma Junfeng

First English Edition 2019
By Royal Collins Publishing Group Inc.
BKM ROYALCOLLINS PUBLISHERS PRIVATE LIMITED
www.royalcollins.com

Headquarters: 550-555 boul. René-Lévesque O Montréal (Québec) H2Z1B1 Canada
India office: 805 Hemkunt House, 8th Floor, Rajendra Place, New Delhi 110 008

ISBN: 978-1-4878-0218-9

We are grateful for the financial assistance of B&R Book Program in the publication of this book.

Broadband China Publishing Project

Steering Committee

Chairman

Shang Bing, Deputy Minister of Industry and Information Technology

Vice President

Cao Shumin, Dean of China Information and Communication Research Institute

Members

Wu Hequan, Scholar at the Chinese Academy of Engineering, Director of the Communication Science and Technology Committee of the Ministry of Industry and Information Technology

Wei Leping, Executive Deputy Director, Communications Science and Technology Committee, Ministry of Industry and Information Technology

Qi Chengyuan, Director of the High Technology Industry Department of the National Development and Reform Commission

Zhang Feng, Director of Communication Development Department, Ministry of Industry and Information Technology

Ao Ran, President of Electronics Industry Press

Editorial Committee

Director

Liu Duo, Associate Dean of China Information and Communication Research Institute

Deputy Director

Jiang Lintao, Director of the Science and Technology Committee of China Information and Communication Research Institute

Committee

Ao Li	Cao Jiguang	Feng Ming
Gao Wei	He Baohong	Li Ting
Liu Jiuru	Luo Zhendong	Tang Xiongyan
Wang Aihua	Wang Chuanchen	Wei Liang
Xu Heyuan	Xu Zhiyuan	Zhao Lisong
Zhang Haiyi		

Editorial Convenors

Wang Xuefei   Wu Ying

Planning Editor

Song Mei

General Introduction 1

China's broadband network is a strategic part of public infrastructure for the nation's economic and social development in the new era. It is an important avenue to drive the modernization of national governance capabilities and the leveling of public services. It is an important channel to facilitate the construction of industrial power, economic development, and new urbanization. The development of broadband networks plays an important supporting role in boosting information consumption and the transformation of the economic development mode, and building a well-off society in an all-round way. It is beneficial for the present and the future to accelerate the construction of broadband networks, enhance capabilities in technological innovation, enrich the application of information services, prosper the development of network culture, and ensure the security of the network.

At present, China has built the world's largest broadband network covering the entire country, and connecting with the world with advanced technology. It also takes a leading role in the world in the number of Internet and mobile phone users, which has led to continued improvement in related industries' capabilities. China has become a true network nation, but on the other hand, its broadband innovative capabilities are relatively underdeveloped. The regional and urban-rural differences are readily evident, and the gap between China's average bandwidth and that of advanced international levels is still quite wide. Further, the network security situation is increasingly dire. Generally speaking, there are many bottlenecks in the development of domestic broadband networks. Faced with the reality that nations around the world are strengthening the development of their broadband strategies and with the rapid development of the ICT industry, China still has a long way to go toward realizing the simultaneous development of industrialization, informationization, urbanization, and agricultural modernization.

The Central Committee of the Communist Party of China and the State Council have placed great emphasis on the development and management of broadband networks. In 2013, the State Council Issued the Broadband China Strategy and

Implementation Plan and the Opinions on Promoting Information Consumption to Expand Domestic Demand. At the end of 2013, the Central Network Security and Informationization Leading Group was established. General Secretary Xi Jinping, personally serving as the team leader, proposed that China should be built up as a strong network nation. Its strategic deployment should be synchronized with the goal of "two centuries" and move toward laying the networking foundation. The basic popularization of facilities, the significant enhancement of independent innovation capabilities, the comprehensive development of the information economy, and the strong commitment to guaranteeing network security are the goals towards which we are continuing to advance. These are the new goals and requirements advanced by the Central Committee for the development of China's broadband network in the new era. These goals require us to uphold the spirit of reform and innovation to ensure security and progress driven by policy, technology, industries, and application. It is important to focus on long-term holistic planning, making great progress through incremental steps, and constantly progressing toward a networking powerhouse.

The Institute of Telecommunications of the Ministry of Industry and Information Technology is an authoritative research institution in the ICT field. For years, it has played an important role in support for major decisions, planning for industry development, providing leadership in setting technical standards, promoting industrial innovation, and providing regulatory support services. The Broadband China Publishing Project series is the crystallization of the knowledge and wisdom of many experts and scholars in the Institute and the industry. It is a concentrated display of multi-disciplinary scientific research achievements as well as a comprehensive integration of years of theoretical and practical experience, aiming to help readers study systematically the latest technologies in broadband networking and capture the latest trends in broadband applications and related industries, thus improving the research, planning, management, and operation of broadband networks. It is our hope that the Chinese government, industry, academia, and research teams will work together to contribute to the development of Broadband China and the construction of a new network powerhouse.

— Shang Bing,
Ministry of Industry and Information Technology

General Introduction 2

Market engine is the driving force for the development of communications. Communication services range from voice to data and video, and the demand for bandwidth is increasing with each passing day. According to a Cisco report, in June 2014, the global Internet traffic at peak hours was 2.66 times the average in 2013. Compared with 2012 figures, average traffic rose by 25%, and busy hours rising by 32%. Cisco also predicted that from 2013 to 2018, global Internet traffic would be 3.22 times previous averages, and average traffic would increase by 23%, while busy hours would increase by 28%. Video has become the mainstream in Internet traffic, and the ratio of global Internet traffic to total usage was projected to increase from 57% in 2013 to 75% in 2018. Global mobile data traffic is growing rapidly, with an 81% increase in 2013 and an average growth rate of 61% by 2018, when mobile traffic was projected to account for 12% of all IP traffic. The international Internet trunk traffic provided by Telegeography in the US rose by an average of 45% from 2009 to 2013, and by 38% in 2013 over 2012 totals. China's international Internet trunk line bandwidth has increased by an average of 39.6%, with 2013 showing a 79% increase over 2012, a growth momentum that is even more obvious.

The development of communication services and technology always complements market engine and technology drive. The integrated circuit continues to follow Moore's Law, and the transistor per unit area has increased by 40% over the past years. The powerful processing and calculation capabilities have improved the spectrum efficiency and signal-to-noise ratio, improved the communication traffic, and better adapted to the growth of Internet traffic. The technological advancement of optional devices coupled with the signal processing of the electrical domain has enabled the commercial capacity of optical fiber communication trunks to be increased by a thousand times per decade. Since 2009, China's mobile communications have moved from 2G to 3G, and on to 4G. With China's advanced multiple access multiplexing and spectrum expansion technology, the peak rate has increased hundreds of times.

In recent years, a noteworthy trend in communication technology and business development is the expansion from consumer applications to enterprise applications.

In 2013, global Internet traffic of enterprises and institutions increased by 21% over 2012, and was projected to reach 2.6 times the 2013 totals by 2018. It would then account for 14% of global Internet traffic, and 14% of global Internet traffic for global enterprise and institutions would be mobile traffic. With the development of the Internet of Things and the deep integration of information and industrialization, Internet applications of enterprises and institutions would see greater development.

The penetration of the Internet has promoted economic recovery. The OECD Internet Economy Outlook 2012, released in 2013, analyzes the impact of the Internet on all sectors of the economy. It concludes that if the broadband penetration rate increases by 1%, GDP will increase by 0.025%. Contributions from the Internet accounted for 4.65% to 7.21% of the GDP in the US in 2010, representing 3-13% of enterprises' added value. The Boston Consulting Group's 2012 Report on Connecting the World analyzes the Internet's contribution to GDP around the world from 2010 to 2016. China ranks third behind the UK and South Korea, and its share of GDP increased from 5.5% in 2010 to 6.9% in 2016. IDC suggests that information technology has evolved from a platform of computers and the Internet to a third platform marked by mobile broadband, cloud services, social applications, and big data, in other words, a broadband platform. It predicts that this platform will generate 40% of the information industry revenue by 2020, and that 98% of growth will be driven by technology on this third platform. The World Bank's research shows that for overseas sales of manufacturing and sales of services, companies using broadband are 6% and 7.5% to 10% higher than other companies. For every ten percentage points of increase in the broadband penetration rate in low- and middle-income countries, their GDP will increase by 1.38 percentage points. The US believes that the development of broadband will drive employment in upstream and downstream industries 1.7 times more than that of traditional industries. According to a research report published by the GSM Association and Deloitte Consulting in 2012, 3G mobile data applications increased by 100%, and per capita GDP growth increased by 1.4 percentage points.

In order to gain new commanding heights of information technology and obtain broadband dividends, some countries have introduced national broadband strategies. In the past two or three years, the US has introduced the National Broadband Plan Research and Development Initiative. Throughout the world, 146 countries have formulated national strategies or plans to accelerate the development of broadband, and many countries have established universal broadband service funds.

The number of Internet users in China is the largest in the world, but the average Internet backbone bandwidth, fixed-line average access rate, and mobile Internet download rate are still lower than the world average. These past years have seen significant improvements, but given the rapid development of the Internet and society and the rising public expectation, it is far from being satisfactory. The State Council issued the Broadband China Strategy and Implementation Plan in August 2013, proposing the initial construction of a broadband, consolidated, secure, and ubiquitous next-generation national infrastructure by 2015, and by 2020, China's next-generation information infrastructure facilities would be relatively close to the levels in developed countries, while technological innovation and industrial competitiveness was projected to reach advanced international levels. The program set out specific development goals, major tasks, and safeguard measures for five areas, including broadband network coverage, network capabilities, application level, industrial chain development, and network information security. It should be expected that the implementation of the Broadband China Strategy will lay a solid network foundation for the development of China's economy and society, bringing great benefits to the public.

As one of the supporting units for the Broadband China Strategy, the Institute of Telecommunications of the Ministry of Industry and Information Technology has conducted in-depth research on the Broadband China Strategy, and it is now launching the Broadband China Publishing Project book series in conjunction with the Electronic Industry Press. This series addresses links, terminals, access, transmutation, network and cloud, covering research, manufacturing, operations and services, involving broadband technology, business, applications, security and management. It interprets the background of the Broadband China Strategy, analyses broadband solutions, and looks forward to the prospects for broadband development. It is a comprehensive and systematic series, reflecting the latest technology of the broadband network and the progress toward international standardization, summarizing practical domestic experiences in China, hence future-oriented and feasible.

Here, I would like to express my heartfelt gratitude to the Institute of Telecommunications of the Ministry of Industry and Information Technology, the Electronics Industry Press, and all the authors for their hard work. I hope this series will help those inside or outside the industry further their understanding of the meaning, connotation, and difficulties of broadband. It is also my hope that this series of books will play a positive role in industry development and government decision-making, contributing positive energy to the implementation of the Broadband China Strategy.

— Wu Hequan,
Director of Communications Science and Technology Committee
of the Ministry of Industry and Information Technology
Chairman of the China Internet Association

Preface

Broadband is a new strategic infrastructure for human society in the 21st century. It is profoundly changing people's production and lifestyle, and has become the cornerstone for strategies formulated by nations around the world aimed at boosting the economy, promoting the transformation of the development mode, creating jobs, and enhancing the long-term competitiveness of those countries. In August 2013, the State Council issued The Broadband China Strategy and Implementation Plan (State Law [2013] No. 31) as a strategic road map and action plan to guide China's broadband development for now and in the future.

The implementation of the Broadband China Strategy aims to strengthen the national top-level design and overall planning, to condense the strength of society, to thoroughly promote the popularization of broadband networks, accelerate the construction of next-generation national information infrastructure, vigorously promote the deepening of broadband in the national economy and social fields, promote in-depth integration of informationization and industrialization, and form the basic platform and important driving force to support economic and social development and technological innovation. The development of next-generation Internet is an important task in implementing the Broadband China Strategy. Accelerating the innovation of next-generation Internet aims to improve independent innovation capabilities and China's information technology, enhance the service capabilities and level of network infrastructure, and enhance the core competitiveness and maintenance of China's information industry. The security of cyberspace is of great significance and is an important starting point for the implementation of the Broadband China Strategy.

With the exhaustion of IPv4 address resources, the global deployment process of IPv6 has been accelerated, and research and experimentation on future network architecture has intensified. The Chinese government has also clearly defined the road map and timetable for the next generation of development and actively promoted the continued evolution of the first generation of the Internet, conscientiously carrying out the scale deployment of the next-generation Internet as it addresses the development needs for the future of the Internet. Active planning is undertaken for innovation and demonstration of the new network's architecture. The next generation of the Internet is the future development direction, involving many new technologies and services. In the process of the development and evolution of the next-generation Internet, more new technologies will continue to emerge.

This book adheres to the principle of "frontier- and practice-orientation, along with foundation." It introduces and compares the typical technical solutions of the future direction of the Internet both in China and abroad. It focuses on the scale deployment requirements of the next generation Internet in China, on the next generation Internet group, network technology, and transition solutions, and introduces the basic technical knowledge of the next-generation and future Internet. The volume provides necessary reference materials for telecom operators, Internet companies, universities and research institutes, technicians to engage in Internet technology research, network construction, operation, and maintenance, and business innovation.

Sections 3.1 to 3.5 of this book were written by Zhao Feng, Chapter 4 by Ma Junfeng, and the rest by Cao Jiguang. In the process of writing this book, we have received the help and support of Gao Wei, Zhu Gang, Zhang Jian, Song Fei, and other colleagues. We have also consulted a large number of domestic and overseas literature in science, and we would like to express our gratitude to all the colleagues and authors.

— the authors

Contents

The Global Development of Broadband and the Evolution of the Next-Generation Internet

Chapter Highlights:
* *The Development of Broadband Around the World*
* *National Broadband Development Strategies*
* *Current Broadband Development in China*
* *China's Broadband Goals*
* *Developing the Next-Generation Internet – a Key Component of the 'Broadband China' Strategy*

Overview

In the 1990s, the IT industry (comprising the twin pillars of telecommunications and electronics) became one of the main driving forces of global economic development. The broadband network is responsible for the high-speed transmission of information, but is more than simply a foundation for information and communication networks such as telecommunications, the Internet, and broadcasting. It has also gradually penetrated politics, economics, culture, finance, education, and healthcare, bringing about major changes to the economy and society more widely.

The development of broadband is propelled by national strategy and guided by relevant national policies. At present, countries and regions with well-developed Internet (such as Europe, America, Japan, and South Korea) have prioritized broadband development by incorporating it into their national strategies. These countries and regions have also rolled out a series of policies to fuse societal demand with network supply and thereby guide the rapid and orderly development of the industrial chain. The hope is to promote the development of broadband so that the whole of society can enjoy its benefits as soon as possible. This is likewise the case in China, where there have been many positive outcomes thus far. However, compared to countries in which the development of broadband is more advanced, the gap is palpable and there remains much room for development.

1.1 The Development of Broadband Around the World

As countries around the world come to a consensus on the immense impact of broadband on both society and the economy, many are rolling out broadband strategies of their own. As broadband access speeds rise, the web environment in individual countries will improve. This will in turn aid the development of sectors such as e-commerce, e-government, and online shopping, which will indirectly promote the development of national economies. Currently, the majority of countries are at the implementation phase of their broadband strategy, characterized by the surfacing of a suite of relevant government policies and available funds. As operators accelerate the construction and deployment of networks to provide faster access, the global broadband market is enjoying strong growth, and the national strategies are starting to bear fruit.

1.1.1 The Current Status of the Global Broadband Market

The global user market for broadband is sizeable. At the end of 2011, a total of 597 million people had broadband access across the globe. Worldwide, the user penetration rate was 10.25%, while the household penetration rate reached 37.75%. The impact of broadband on economic development is increasing, and it is the key to gaining an advantage in the new technological revolution.

1) A rapidly-growing global user base with vast potential

At the start of 2010, the number of broadband users around the world entered a new phase of accelerated growth. In both 2010 and 2011, the annual increase in users exceeded that of the preceding year. Several years on, the growth of broadband user numbers globally is set to continue. It was estimated that the average increase in new users between 2012 and 2016 would be in excess of 80 million per annum. In 2016, the total number of broadband users worldwide is expected to exceed one billion, with user penetration surpassing 15%. See Figure 1-1 below.

2) Rapidly rising market share of broadband users in developing countries (but lagging behind developed countries in other aspects)

The market share of broadband users in developing countries increased from 44% in

2006 to 62%, but they lag far behind developed countries for penetration and pricing.

In terms of penetration, the rate in developed regions such as Western Europe and North America is in excess of 30%, while in developing countries and regions, the rate still hovers around 10%. See Figure 1-2 for details.

In terms of pricing, according to data from the ITU, the cost of entry for broadband services in developed countries and regions is less than 0.01 of the per capita GNI, while in developing countries and regions, it rises to more than 0.05 of the per capital GNI.

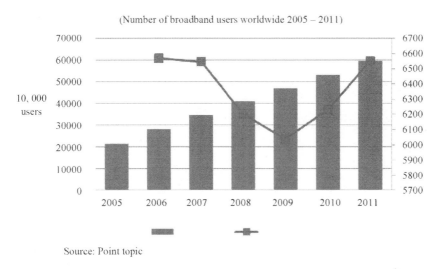

Source: Point topic

Figure 1-1: Number of broadband users worldwide, both accumulated and new

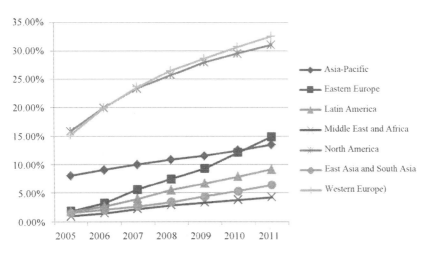

Figure 1-2: Broadband penetration in various regions between 2005 and 2011

3) A global increase in internet access speeds

Large-scale construction of broadband network infrastructure and the increasing use of fiber to the x (FTTx) in broadband network architecture means that broadband internet access speeds worldwide are rising significantly. According to a monitoring report by Akamai, the global average internet access speed in the fourth quarter of 2011 increased by 19% year-on-year to reach 2.3 Mbps, with an average peak access speed of 11.7 Mbps. In the fourth quarter of 2011, the global high-speed broadband (>5 Mbps) usage rate was 27% - an increase of 17% year-on-year. However, the usage rate of regular-speed broadband (greater than 2 Mbps) was 66% - an increase of 9.2% year-on-year. The global usage rate for narrowband internet access (less than 256 kbps) continued to fall to 2.5%

4) Rapid replacement of copper cable narrowband with fiber-optic broadband

The increase in demand arising from the development of high-end businesses and the transmission of high definition (HD) video streams means that the replacement of ADSL with VDSL and PON for network access applications will continue to accelerate. The same goes for data transfer speeds.

Around the world, income derived from fiber-optic access equipment has grown relatively quickly. In 2011, the figure was US$2.1 billion accompanied by a large increase in market share from 18% in 2008 to nearly 30% in 2011. This ranks fiber-optic access second in terms of market share, and it is poised to replace DSL as the leader.

5) Rapid development of mobile broadband

In recent years, mobile broadband access has seen rapid development as countries around the world have made use of a variety of access methods to provide users with seamless connectivity. According to statistics from the ITU, by the end of 2011 there were nearly 1.2 billion mobile broadband users, with annual growth of 45% over the previous four years. The result is that for every fixed broadband user, there are two mobile broadband users.

1.1.2 Global development goals

The international community understands the importance of broadband and is intensifying its efforts to promote it. In October 2011, the United Nation's Broadband Commission for Digital Development confirmed four new international broadband development goals to be achieved by 2015.

- Formulation of a broadband policy. By 2015, all countries are to have a national broadband plan or strategy, or are to include broadband access in their definitions of ordinary access/service.
- Lowering the threshold for broadband adoption. By 2015, all developing countries are to ensure that prices for entry-level broadband services reach an acceptable level through reasonable controls and market regulation. For example, broadband expenditure should be less than 5% of the average monthly income per capita.
- Increase the proportion of households with broadband connectivity. By 2015, 40% of households in developing countries and regions should have Internet access.
- Encourage residents to go online. By 2015, the penetration rate of global Internet users should reach 60%, with a penetration rate of 50% in developing countries, and 15% in the least developed countries (LDC).

At the beginning of 2012, the United Nation's Broadband Commission for Digital Development called for broadband to be included in the three pillars of sustainable development, namely the economy, the environment, and society. The aim was to use broadband to promote the development of a sustainable future with a low-carbon footprint.

1) The formulation of national broadband strategies and an analysis of the goals

Countries are keen to make the most of this historical opportunity for broadband development by formulating and then implementing a national broadband strategy and action plan. Under this initiative, to which the international community has attached great importance, some 110 countries have rolled out plans, such as 'Connect America', 'Digital Britain', 'Digital France', 'Smart Nation Singapore', and 'i-Japan'. The focus is on making broadband more accessible, improving access speeds, and increasing the number of applications. At present, most of the countries and economic bodies that have announced a broadband strategy are clustered in Europe, totaling 37, or 33.6% of the global total. This is followed by countries and economic bodies in Asia Pacific, with 21.8% of the global total. See Figure 1-3 for details.

The main goals in these broadband strategies are:

- To achieve stronger network infrastructure and faster access speeds (i.e. availability);

- To provide broadband access across the board (accessibility, affordability), improve coverage;
- To realize socio-economic development through broadband applications.

The focus of each country's broadband strategy is on the following three areas:

(1) Accelerating the construction of an ultra-high-speed broadband infrastructure
Increasing infrastructure capacity lies at the heart of the broadband strategy. Looking at the development strategies of various countries around the world, the primary aim is to come up with construction goals for high-speed national broadband infrastructure based on the country's economic development, existing broadband infrastructure, and broadband needs.

In Japan and South Korea – the two countries that are the world's leaders in terms of broadband access – 100 Mbps access is commonplace. The plan is to provide 1 Gbps access speeds within the next two to five years. Additionally, Singapore and Japan hope to achieve ultra-high-speed broadband household penetration rates of more than 90%.

In Western and Northern Europe, where broadband access is more advanced, universal coverage has mostly been achieved. Therefore, national broadband development goals are to provide access speeds of 25–100 Mbps within the next five years or so, and household penetration rates of 75–100%. In America, the number of users in each of the three internet access categories (256 Kbps–2 Mbps, 2–10 Mbps, and more than 10 Mbps) is roughly the same. The long-term goal is to provide 100 Mbps access speeds to 100 million households.

Many developing countries have also rolled out plans to increase broadband speeds that are commensurate with domestic conditions. For example, India made plans to

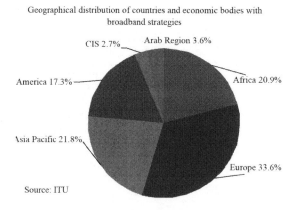

Geographical distribution of countries and economic bodies with broadband strategies

CIS 2.7%   Arab Region 3.6%

America 17.3%   Africa 20.9%

Asia Pacific 21.8%

Europe 33.6%

Source: ITU

Figure 1-3: Geographical distribution of countries and economic bodies with broadband strategies (as of February 2012)

realize 10 Mbps access speeds for households in big cities, 4 Mbps access speeds for households in small to medium-sized cities, and 2 Mbps access speeds for towns and rural areas in 2014.

(2) Narrowing the digital gap to provide broadband access for all

Governments worldwide are using new measures to improve access to broadband services. The goal of many developed countries is to provide all-round broadband services within their borders. In developing countries where the broadband penetration rate is lower, governments have come up with goals tailored to the present situation in order to improve coverage in villages and rural areas.

Currently, more than 20% of countries around the world have embarked on efforts to improve access to broadband services. The United Nations has recommended that all countries include broadband in their definition of regular internet access by 2015. In its 'Broadband Canada-Connecting Rural Residents' plan, Canada invested US$225 million into broadband infrastructure in rural areas, providing suppliers of broadband access facilities with once-off national financial support. In 2010, the government of Brazil officially unveiled the 'Brazilian National Broadband Plan'. The goal was to provide broadband coverage to 4,000 medium to small cities by 2014. In so doing, 40 million households would have broadband access of at least 1 Mbps or even higher, raising the broadband penetration rate to 68%. At the same time, monthly fees would be lowered to between 15 to 35 Brazilian Lira for optimum affordability. In terms of the development of fiber-optic networks in rural areas, India planned to invest US$4.18 billion through its ordinary services fund, with an additional US$3.09 billion coming from the 'Mahatma Gandhi National Rural Employment Guarantee Scheme'.

(3) Widening and deepening broadband business applications for economic and social development.

Countries around the world have designated the application of broadband in economic and social development as an important goal. Efforts are being made to develop new services and cultural industries based on broadband, and to raise usage levels among individuals as well as small and medium-sized enterprises.

America is making considerable efforts to put broadband to use in healthcare, education, energy, the economy, governmental services, citizen participation, and public and national security in order to achieve its national target. By 2015, the European Union aims to raise internet use from 60% to 75%, lift the proportion of online shoppers to 50%, and encourage 33% of small and medium-sized enterprises to launch e-Businesses. In addition, the European Commission has activated the 'European eGovernment Management Action Plan' for the next five years. The plan consists of 40 measures to allow citizens and companies to complete a series of government-related activities online, such as registering a company, applying for welfare benefits or health insurance, registering for university studies, and enterprise bidding. In developing countries such as Cuba, Peru, Chile, and Argentina, plans for online undergraduate courses have been implemented, building virtual universities to promote electronic networking within higher education.

2) The implementation of broadband strategies

Countries that rolled out their broadband strategies early on (such as South Korea, Japan, Finland, and Sweden) have already reached certain milestones. For example, Sweden's goal to provide broadband speeds of at least 100 Mbps to 40% of households and companies by 2015 was achieved four years in advance. In South Korea, 100 Mbps broadband access was commonplace as early as 2007, and Japan's 'e-Japan' plan was also realized ahead of schedule.

Most countries are in the implementation phase of their broadband strategies.

- America: Most of the US$7.2 billion government broadband funds are in place, and construction is expected to be completed by September 2013.
- Australia: In September 2011, the NBN completed Phase One of its plan to provide fiber-optic network coverage to five regions. Phase Two is currently underway. At the same time, new plans and government incentive policies are constantly being rolled out.
- The European Union: A large-scale investment plan amounting to 50 billion Euros was launched in October 2011. The funds are earmarked for the development of transportation, energy, and broadband networks to ensure sustained growth and increase employment rates.
- The United Kingdom: The government has provided one billion British Pounds in funding for the nation's broadband plan, and it is gradually being implemented.

In the current challenging economic environment, broadband facilities and services can raise a country's competitiveness, promote social and economic growth, and create job opportunities. For both developed and developing countries around the world, expanding broadband access facilities and services are critical development strategies.

3) Analysis of ultra-high-speed broadband applications overseas

The development of broadband information technology implies immense potential for economic growth. It can result in support for higher-quality internet services, smarter government services, and telecommuting. This in turn can create new areas of growth in the economy, thereby stimulating economic development. A high-speed broadband network can support applications in the following areas:

- Providing high-quality interactive broadband applications such as video-conferencing, home video surveillance, streaming media and HDTV;
- Promoting the use of information technology in traditional industries, such as e-Government, e-Commerce, and e-Learning.

Presently, ultra-high-speed networks in world-leading countries (such as South Korea and Japan) aim to help create seamless and omnipresent internet coverage that allow users to enjoy the benefits of broadband anywhere, anytime.

(1) South Korea

South Korea's goal for the development of broadband applications is to build an 'always connected' society. IT is used to create convenience for residents in the areas of dining, shopping, housing, travelling, sports, and entertainment. Some of the major broadband applications on which the government is focusing include smart transport, smart cities, digital households, and an emergency information system. The South Korean government has also targeted the public sector by providing highly-affordable internet services, as well as free websites for schools. This has been done to accelerate the digitization of society and to improve science and technology as well as education.

Broadband is already a part of every South Korean's life, and the internet is used to solve all sorts of day-to-day problems. South Korean internet users are transiting from using the internet passively (to obtain information) to more interactive activities such as online polling, shopping, gaming, learning, and remote-control of household appliances via mobile phones. At present, nearly a quarter of retail business is carried out online. Internet banking and online investment in stocks are also very common. The development of online gaming in South Korea is also seeing a flurry of activity. In terms of PC gaming expenditure per capita, South Korea ranks first globally. Annual revenue in the country's online gaming industry is growing at an average of 9.7% year-on-year, and is expected to reach US$5 billion in 2016.

(2) Japan

Japan's plan is to make use of ICT to reshape society and create a large-scale platform to provide a variety of IT applications. This would allow applications for daily life to be developed and rolled out at lower costs and higher speeds. The overall development goals of the three main areas under the 'i-Japan' strategy are as follows:

- e-Government: Promoting administrative reform based on digital technology to raise convenience. Realizing government services that are streamlined, highly efficient, and standardized, which can be "visualized" by 2015.
- e-Healthcare: By 2015, digital technology and information will make vital contributions to some of the problems arising from an ageing population, lack of doctors, and regional imbalances. This will improve the quality of healthcare.
- e-Learning: By 2015, information technology will be used to cultivate human resources in kindergartens, nursery schools, primary and secondary schools, and universities. This includes promoting digital technology applications, interactive learning and teaching, as well as raising students' ability to make use of digital information.

The Japanese government also recognizes that in the era of cloud computing, all sorts of information and services can be made available online. Through implementing the 'i-Japan' strategy, the government hopes to develop new industries that can support the Japanese economy in the medium to long term. It is also dedicated to embarking on major projects such as a smart transportation system and environmentally-friendly technology with green IT at its core.

1.1.3 A Global Broadband Development Model

After several years of development, most countries have settled on development models that are mainly made up of corporate investments and market regulation mechanisms. However, this alone cannot fully realize a country's broadband development goals. Therefore, countries are generally relying on methods such as injecting public funds to promote broadband development on a national level. In this way, the models can be grouped into four categories: the direct government investment incorporation model, the Public-Private Partnership (PPP) model, the Private Finance Initiative (PFI) model, and the government-subsidies model.

(1) Government investment incorporation model – highest government involvement
In this model, the government invests in setting up a company that is tasked with constructing a brand-new national broadband network. This model is the one with the most government involvement, that is to say, the government is directly involved and has the most say. It is also the model in which public money accounts for the biggest proportion of the funds that are injected. Australia, Brazil, Qatar, and Rwanda are classic examples of countries using this model.

Often, countries that choose this model do not have a nationwide broadband network, or the development of such a network is severely lagging behind those of other countries. Additionally, the country is determined to develop broadband and significantly raise the capabilities of its broadband network. Prior to rolling out its broadband network development strategy, Australia was lagging behind other developed countries, especially in the construction of fiber-optic networks, slow network speeds, and high usage fees. There were two main reasons for this. First, individual state parliaments had different policies for broadband development, causing operators to slow the pace of broadband construction in order to minimize policy risks.

Second, Australia has a huge land mass and is sparsely populated. This means that the costs of building a broadband network are high. With profit in mind, operators are generally not keen to invest heavily in broadband construction. To a certain extent, the current state of Australia's broadband network has curtailed its economic vibrancy. As a result, it has set ambitious goals such as connecting 93% of households, schools, and workplaces using fiber-optics by 2020, and providing maximum transmission speeds in excess of 1 Gbps. To achieve its strategic goals despite its unique geographical conditions, Australia's government decided to make use of public funds to set up a company tasked with constructing a brand-new national fiber-optic broadband network.

Under this model the government has a high degree of control, but it does not replace market regulation mechanisms. The government's investment is primarily for the construction of a nationwide backbone network, while the private sector is still responsible for providing access to it. Market mechanisms still playing a role in the regulation of network access and operation. Through the wholesaling of the broadband network, operators obtain usage rights for the government-constructed backbone network. The Australian government also has an exit plan in place for its investment, to ensure that market regulation mechanisms can take the lead. It has stipulated that it will sell the majority of its shares five years after the network is fully operational, to allow the market regulation mechanisms to fulfill their functions.

The advantage of this model is that it allows an open market for the network to be easily implemented, and helps foster a fair and competitive environment. Since the backbone network is constructed through government investments, there are considerably fewer barriers to opening up the network for access to all as compared to backbone networks that are built using private funds. Additionally, since the government and the network operators do not have many overlapping interests, it is very easy to create a competitive environment that is fair to all.

The most significant drawback of this model is the large amount of public funding required. The Australian government has invested $27.5 billion Australian dollars into the development of broadband, which is far higher than that of countries or regions that are using other models. Excessive investment can exert huge fiscal pressure on the government.

(2) PPP model – sharing funds, risks, technologies, and experience

In the PPP model, the government and enterprises co-invest in constructing the broadband network. What sets this apart from the government investment company model is that the emphasis is on the government and enterprises making joint decisions, sharing technologies, and splitting the operational risks. Additionally, under the PPP model, the public funds injected by the government can generate gains through the equity held. The amount of investment needed from the government under the PPP model is smaller than that of a government-investment company. The model is also adaptable, and allows the government to share the risks with enterprises. As such, it is being utilized by an increasing number of countries, including most of Europe.

The main reason countries use the PPP model is to provide universal broadband services domestically. In Finland's broadband development strategy, the goal for normal services is to ensure that more than 99% of the population is within 2 km of a 100 Mbps fiber-optic or cable by 2015. If Finland were to rely solely on investments from enterprises and market regulation mechanisms, then only 94% of its population would be covered by 2015. Broadband network coverage for the remaining 5% in remote areas can only be realized through an injection of public funds. In order to provide normal service and for more efficient use of public funds, the Finnish government stipulated that the PPP model will only be used, and public funds injected when corporate financial evaluation indicates that the undertaking is 100% infeasible.

Under the PPP model, market regulation mechanisms play a leading role. Normally, countries restrict the proportion of total investments that can come from public funds. In Finland, the government limited this fraction to no more than a third when public funds were used in the PPP model to construct broadband networks. The European Union and the municipal government accounted for another third, with the remaining third coming from enterprises. The Spanish government stated that it would invest 84 million Euros in public funds with 31 million Euros coming from the European Union's structural funds, and the remaining 53 million Euros in the form of interest-free government loans. On the other hand, operators would invest 280 million Euros.

(3) PFI model – primarily for public use

Under the PFI model, the government first provides enterprises with the service requirements of broadband usage. The companies are then responsible for constructing the network. Finally, the government pays the enterprises for broadband services. Through this model, the government funds the enterprises indirectly. The main difference between this model and the previous two is the role the government plays. Under the first two models, the government is responsible for building the broadband network, while in the PFI model, it is a buyer of broadband services.

The PFI model is primarily for broadband application. Brazil is recognized as the most successful example. The government prioritized the use of broadband in sectors such as the government, education, and public access. Enterprises are responsible for laying the broadband access points. Upon completion, the government buys the services before providing free broadband access services to the public sector, public schools, libraries, and the general public.

Some European countries, such as the Netherlands and Italy, make use of the PFI model to provide broadband access services in remote areas. The returns on investments in broadband network construction in remote regions are low. As such, operators are not interested in investing even though the demand for access in these areas are growing. In view of this, the government uses the PFI model to develop broadband services.

Overall, the PFI model plays a vital role in raising the use of broadband services, bridging the digital divide, and stimulating economic growth.

(4) Government-subsidies model – an effective supplement to private investment
Within this model, the government also provides the enterprises with a sum of money. What sets this model apart from the government-owned company and PPP models is that the enterprises do not need to repay the funds received. The government is also not involved in the enterprises' decision-making process. In the course of promoting the development of broadband on a national level, the vast majority of countries worldwide can be said to have utilized this model to varying degrees, including primary use by the USA and Japan.

This model is also used to provide universal access to broadband services. For example, in the United Kingdom, BT has committed to constructing a fiber-optic network. However, solely relying on BT's investments would mean that only two-thirds of households in UK would have access to the fiber-optic broadband network by 2015, falling short of the government's target of coverage for all.

This model emphasizes public funds as an effective supplement to private investment, not a substitute. Even though there is no need for the enterprises to repay the subsidies given, the government would often require that the network infrastructure be made available to all, i.e. even competitors can have fair and equal access to the facilities.

Through an analysis of the four broadband development models mentioned above, and having researched the main national broadband development models, we wish to highlight the following:

- The injection of public funds is not designed to replace private investments and market regulation mechanisms. Even in the government-owned company model, which has the greatest amount of governmental involvement, only in certain segments (i.e. the backbone network) do government controls supersede market regulation mechanisms. Moreover, there are exit mechanisms in place for the government's public funds so that market regulation mechanisms can fulfil their leading roles. In the PPP model, the degree of governmental control is restricted by limits placed on the proportion of funds provided by the government. In both the PFI and government-subsidies model, the government controls broadband development indirectly.

- Using a combination of broadband development models in order to meet a country's developmental needs. Each broadband development model has its strengths and weaknesses, and there is no single model that can meet all of a country's developmental needs. As such, countries around the world generally make use of a combination of models concurrently. For example, the majority of European countries make use of a combination of the PPP, PFI, and government-subsidies models.

- Selecting an economically viable and appropriate broadband development model. In drawing up broadband development goals and models, countries tend to weigh up major issues such as the long-term economic and societal returns of broadband development, calculate the direct and indirect benefits and effects of broadband on the country, and assess the direct and indirect costs arising. An economic evaluation is also carried out on broadband investment, both public and private. It is on this basis that economically viable and appropriate broadband developmental goals and models are decided.

1.1.4 Government policies that support broadband development

Foreign governments are playing an increasingly important role in broadband development, to establish an ecosystem that addresses both supply and demand. These governments support the 'supply' of broadband networks and businesses (especially in regions that are not economically viable) by regulating the market, expanding the coverage of universal broadband service, and having a flexible licensing system. They are also investing directly in network infrastructure and facilities, providing a larger spectrum for use, tackling bottlenecks in network construction and use, and rolling out market-friendly tax policies. At the same time, these governments are stimulating broadband demand and use through raising public awareness and technological knowledge, subsidizing certain user segments, and rolling out more public services.

1) 'Supply' – the main policy for upgrading broadband network infrastructure

A government has three main tasks in this regard:

- To create a level playing field
  Practical experience has shown that effective competition can promote the development of broadband, improve the quality of network infrastructure, and raise the efficiency of use. Competition is especially important for lowering costs and improving service quality. As such, governments need to prioritize policies that aim to create a level playing field.

- To encourage corporate investment
  On the one hand, experience has shown that corporate investment plays a leading role in promoting broadband development. On the other hand, the cost of government investment is high, and funds are limited. Therefore, it is impossible to rely solely on government investment. The government needs to consider how to guide and encourage corporate investment.

- To create a mechanism for universal broadband services
  Relying only on the market to regulate the investment and construction of broadband networks is insufficient. This is especially so in regions where geographical conditions are poor and the population density is low. As the return on investment for building broadband networks in these areas is also low, enterprises are reluctant to invest. As such, government intervention is needed in these regions to ensure the realization of universal access to broadband services.

(1) To create a level playing field with open network access

The main policy adopted by various countries to ensure an even playing field is open network access. In terms of regulation, the consensus is to ensure that there is open access to broadband infrastructure. The scale and scope of the investment involved in a broadband network can be undertaken by a leading operator. If access to the broadband infrastructure is not open, a monopoly of the broadband market can result, which would impede healthy competition.

Countries around the world differ in how they provide open access to their broadband infrastructure networks. In Australia, the government has set up a company of its own, which is responsible for constructing a national backbone broadband network from scratch. This network is then made available to all broadband service providers without discrimination. However, in Singapore and the United Kingdom, the broadband network is split according to function.

The current consensus is that there needs to be open access to the broadband network's infrastructure layer at the physical, data link, and network levels. This layer is uniquely important due to its non-replicability. Regulatory supervision that ensures open, fair, and transparent access to the infrastructure layer would be beneficial for promoting fair competition across the broadband sector. As for the transport layer, there is no clear consensus on whether open access is necessary. Finally, few believe that open access is required at the application and service layer (comprising the session layer, presentation layer, and application layer).

(2) Policies to attract private investment
- Provide low interest loans
  Low interest loans can reduce the financing costs of corporate investments, thereby encouraging such investments in the broadband sector. The Japanese government requires the Development Bank of Japan to provide low-interest loans to domestic telecommunication operators for the purpose of developing broadband access services. To develop a broadband network in rural areas, the South Korean government has provided low-interest loans worth US$770 billion to KT.

- Tax relief for broadband-related industries and accelerated depreciation of assets
  Some countries are looking at broadband-related tax relief to encourage corporate investment in the sector. The Portuguese government is considering policies such as tax rebates to stimulate the construction of a broadband infrastructure. To make broadband services more affordable, regulatory authorities in India have requested that imported equipment be exempted from tax, and have asked for accelerated depreciation of assets. The Japanese government is providing broadband access operators with tax incentives such as corporate tax redemption, as well as the accelerated depreciation and amortization of fixed assets.

- A wireless broadband spectrum
  As the industry evolves towards more powerful and seamless broadband everywhere, the demand for a wireless broadband spectrum increases exponentially. Providing enterprises with the resources they need to create a wireless broadband spectrum will also encourage them to invest in the sector. An increasing number of countries are coming up with plans to make better use of their wireless broadband spectrums. First, future demands are calculated. Next, more frequency bands are made available to enterprises through measures such as the accelerated release of invalid and

inefficient spectrums, and spectrum sharing. At the same time, research into the 'digital dividend' and White Space are being sped up in search of higher quality wireless broadband spectrum for enterprises.

- Providing open access to infrastructure
  Civil engineering accounts for the bulk of the construction cost for wired network infrastructure. Providing access to public infrastructure such as roads, railway, pipelines, and the electrical grid can significantly lower civil engineering costs, thereby stimulating greater corporate interest in broadband network investment. There are three ways in which this can be achieved. First is to lower access costs to a level that is acceptable to broadband network builders. Second is to streamline the legal process for obtaining access. Third is to provide free access to broadband network builders for infrastructure that is controlled by the state.

(3) A universal service mechanism

In the past, most countries defined universal service as prioritizing the demand for fixed line telephony. Currently, some nations are including broadband in the domain of universal service through setting up new funds and modifying existing ones. Universal service funds as a long-term protection mechanism is used to support the development of universal broadband services in rural and less-developed areas. A levy is imposed on the operators' revenue according to a pre-determined ratio. The regulatory agency then redistributes the sum collected to operators that provide services in areas with high costs, such as rural and remote regions. Some of the money is used to subsidize projects that help low-income groups and school libraries connect to broadband networks.

In America for example, the 'Connect America Fund' (CAF) will be set up to provide the general public with affordable broadband services at download speeds of at least 4 Mbps. The FCC also has plans to increase the Universal Service Fund (USF) from US$8 billion to US$15.5 billion within the next decade to support broadband construction. If the United States Congress wishes to accelerate the realization of universal broadband service, it can provide a few billion dollars' worth of public funds annually over the next two to three years.

2) Demand - Policies for Enhancing Broadband Demand and Application

These policies are closely related to a country's cultural and economic background, as well as to the status of its broadband development. Therefore, there are significant differences in specific national policies. Currently, it is mostly developed countries that have them. The value of such policies to developing countries would depend on the actual circumstances.

At present, the policies for enhancing broadband demand and application can be grouped into three categories.

(1) Network side
- To provide schools with broadband access;
- To prioritize governmental usage;
- To use the USF to provide network coverage to communities that currently have no broadband access;
- To establish a broadband access center the communities; and
- To designate broadband access as a universal service.

(2) Business and application side
- To develop government-led incubators for broadband application;
- To provide e-government broadband application;
- To promote the development of digital content, and to support local content;
- To promote broadband application in the healthcare, educational, and agricultural sectors.

(3) User side
- To provide educational institutions with low-cost user terminals;
- To boost digital literacy;
- To make broadband access affordable for the end-user;
- To monitor service quality;
- To create a secure environment for e-commerce;
- To offer training for small and medium-sized enterprises.

An analysis of these policies indicates that there are three major problems on the demand side.

- *Underlying issues are hindering more widespread adoption of broadband*
  If the potential consumer group in a country that is receptive to broadband services has become an actual consumer of broadband services, then the main challenge the country faces in raising the demand for broadband is how to encourage adoption among individuals who do not understand it or are less interested. The general measures taken by various countries in this regard are: providing these potential consumers with more information about broadband,

and improving digital literacy; and encouraging the application of broadband in education as well as in small and medium-sized enterprises.

- *Provision of affordable broadband services*
  If a country's broadband tariffs are too expensive for the majority of people, its government has a role to play in providing subsidies and terminals to lower costs for the user.

- *Attractive broadband applications*
  If a country is already at a relatively advanced phase in its broadband development, it must focus on improving the standard of its cross-broadband applications and enhancing innovation.

Additionally, there is value in the government policies put in place to support the enhancement of broadband applications. Such policies can be grouped into two categories:

- *Government-built platforms for public services, education, and healthcare based on broadband*
  In order to promote broadband applications, some governments have invested in public services, education, and healthcare platforms, for example the €50 billion "Connecting European Infrastructure" project that was approved in October 2011. Part of the €50 billion was earmarked for investment in broadband infrastructure, of which €9.2 billion was designated for the construction of the infrastructure needed for e-identity cards, e-procurement, e-health records, European digital libraries, and e-justice and customs services.

- *Resolving issues such as connection fees, end-user terminals, and computer skills.*
  In foreign countries, part of the universal service fund is used to subsidize broadband connection fees for those facing economic hardship. In the past, the subsidies were used for telephony services, but they are gradually being directed towards broadband services instead.

End-user terminals are the basis for the smooth implementation of universal broadband services. At present, end-user equipment also includes new equipment such as smart phones and netbooks in addition to personal computers. In order to provide universal broadband services, it is necessary to provide end-user terminals for the economically disadvantaged, and for institutions (such as primary and secondary schools) that receive financial aid from the government. In addition to building networks and deploying end-user terminals, there is a need to develop user awareness and ICT skills, and improve digital literacy. For example, the United States has set up a national digital literacy group that organizes training for young people and adults, and teaches digital skills to narrow the education gap.

The E-rate project in the United States provides universal broadband service to schools and libraries. According to suggestions from the broadband initiative, the FCC must update and upgrade it. Presently, 97% of schools and almost all public libraries in the USA have basic internet access, but connection speeds are slow. The latest E-rate policies approved by the FCC include:

· High-speed fiber – Funding for schools to obtain fiber access. There are several options, such as existing regional and local networks, or using unutilized local, high-speed fiber-optic network routes;
· Hotspots in schools – Schools can set up hotspots to connect the surrounding community, in order to offer students Internet access at home, and at the same time promote the development of the local area;
· Learning On the Go – This FCC pilot uses wireless end-user terminals such as netbooks and tablets both inside and outside the classroom, eliminating the need for students to learn in a fixed location.

3) State Investment – Providing Financial and Tax Policy Support for Broadband Development

Foreign governments generally provide financial and tax support for the development of broadband. There are differences in the magnitude, manner, and areas in which support is rendered.

(1) Magnitude of support by the state

The amount and intensity that countries put into developing broadband differ according to their objectives, current development level, financial resources, and intended mode of support, as shown in Figure 1-4.

In terms of amount invested, Australia takes the lead with AU$27.5 billion.

In terms of per capita public investment, New Zealand ranks first with US$200 per capita. This is followed by Australia with US$160 per capita. See Figure 1-5 for specifics.

(2) Mode of support

Foreign governments use various means to support broadband development, mainly coming under the categories of direct and indirect support.

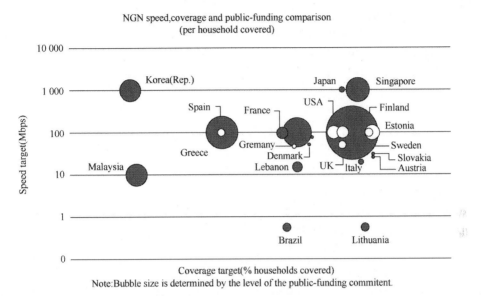

Figure 1-4: A comparison of the magnitude of support from various countries

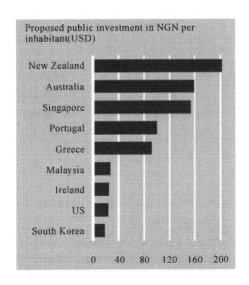

Figure 1-5: Public investment per capita

- Direct support
  - A one-off injection of funds (see Table 1-1)
  - Improving the existing universal service mechanism for broadband development In the past, most countries defined universal service as giving priority to fixed-line telephony services. Currently, some countries have included broadband in the universal service category by setting up new funds or revising existing universal service funds. A universal service fund is a long-term guarantee of support for the construction of universal broadband services in rural and underdeveloped areas.

Table 1-1: One-off injections of funds by various countries

United States	Part of the US$7.2 billion CAF in the economic recovery plan is from direct grants
Australia	Direct investment of AU$27.5 billion
South Korea	In February 2009, South Korea announced that its central government would directly invest 1.3 trillion Korean won in broadband construction
Brazil	The Brazilian government announced a total investment of 11 billion riyals (about US$6.1 billion) in the PNBL (Plano Nacional de Banda Larga or National Broadband Plan). This includes the 3.22 billion riyals (US$2.03 billion) invested in Telebras, the former state-owned telecommunications operator, in the four years since 2011.
United Kingdom	The government plans to directly invest £830 million
Finland	The government intends to directly invest 67 million Euros

Table 1-2: Status of improvements in various countries

United States	The CAF (Connect America Fund) will be set up so that the general public is can enjoy affordable broadband services with actual download speeds of at least 4 Mbps. The FCC plans to increase the Universal Service Fund (USF) from US$8 billion to US$15.5 billion within the next 10 Years to support broadband development.
Brazil	The national telecommunications investment fund (Funntel) will provide 1.75 billion riyals for related research and development projects to launch its broadband program.
South Korea	In February 2009, South Korea announced that its central government would directly invest 1.3 trillion Korean won in broadband construction.
Brazil	The Brazilian government announced a total investment of 11 billion riyals (about US$6.1 billion) in the PNBL (National Broadband Plan). This includes the 3.22 billion riyals (US$2.03 billion) it invested in Telebras, the former state-owned telecommunications operator, in the four years since 2011.

- Indirect support
  - Tax relief and accelerated asset depreciation (see Table 1-3)
  - Low-interest loans (see Table 1-4)
  - Areas of support
    The areas in which countries provide support for broadband can be divided into supply and demand. Supply refers to the construction of broadband infrastructure while demand covers the development of broadband applications amongst users. Government policies that support broadband development in various countries mainly target the supply side.

    a. Supply
    Focuses on building broadband networks in remote areas such as villages (see Table 1-5).

    b. Demand (see Table 1-6)

Table 1-3: Tax relief and accelerated asset depreciation in various countries

Japan	The government provides tax benefits to broadband access operators, including corporate tax redemption, and tax incentives on the depreciation and amortization of fixed assets
India	India exempts ISPs (internet service providers) from income tax for the first five years of business. The original tax rate is 8% of the ISP's business revenue; request for duty-free asset import
Portugal	The government has pledged to enact laws to stimulate broadband development through tax relief
Brazil	Divided into tax-exempt states and non-tax-exempt states. High-speed Internet access of at least 1 Mbps is available at a price of 29.80 riyals/month in tax-exempt states and 35 riyals (US$22)/month in non-tax-exempt states.
United Kingdom	Two thirds of the funds for the construction of broadband networks come from private enterprises, and the capital is exempted from tax in the first year

Table 1-4: Low interest loans in various countries

Japan	The Development Bank of Japan provides debt guarantees for broadband access, and low interest rate financing
South Korea	The government has provided KT (Korea Telecom) with a low interest loan of US$77 million in order to build broadband networks in rural areas
Brazil	The national development bank (BNDES) will provide Telebras and other operators with 7.5 billion riyals in low-interest loans

Table 1-5: Support on the supply side

United States	The US$7.2 billion CAF is mainly used to expand the coverage of broadband access, thereby increasing the broadband penetration rate
The European Union	Key investment of 1 billion Euros to develop Internet infrastructure in the remote areas of EU member states
United Kingdom	The government-invested Next Generation Fund is mainly used for the construction of next-generation networks in a third of the UK (relatively remote areas)
India	Only US$4.18 billion was invested through universal service for the construction of fiber-optic networks in rural areas

Table 1-6: Support on the demand side

United States	Funding will be provided for computer centers in libraries, universities, and other public-sector units, along with projects to train students to use the Internet. There is also a US$350 million fund set aside for the broadband roadmap.
United Kingdom	US$480 million worth of broadband subsidies have been provided to low-income families

Table 1-7: Typical fiscal policies for broadband development

United States	US$7.2 billion broadband fund in the economic recovery plan.

This US$7.2 billion includes grants, loans, and loan guarantees. Of the total, US$4.7 billion is provided by the US Department of Commerce and the National Telecommunications and Information Administration (NTIA) through BTOP (the Broadband Technologies Opportunities Program). The remaining US$2.5 billion is provided by the Rural Utilities Services (RUS) of the US Department of Agriculture (USDA) through BIP (the Broadband Initiatives Program). All these funds will be allocated in accordance with the requirements of the economic recovery plan to expand broadband coverage and increase the penetration of broadband networks.

The US Department of Commerce projects tend to take a "government grants + (funding) package" approach. The focus is on the so-called "middle mile" projects linking libraries, universities, and public security agencies. The department also provides funding for computer centers and Internet training courses in libraries, universities, and other public entities. Another US$350 million is set aside for the broadband map program. In total, more than 200 projects are being funded, most of which are for network construction. See the figure below for a detailed breakdown.

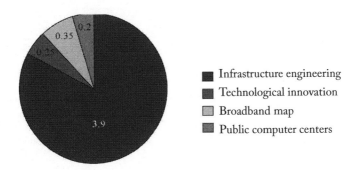

■ Infrastructure engineering
■ Technological innovation
▨ Broadband map
■ Public computer centers

Examples of network projects include the 260-mile fiber loop project in the mountainous region of northern Georgia (US$33.5 million from the government and a US$88 million from the North Georgia Network Cooperative) and the fiber-optic network in the mountainous regions of Maine. The computer centers project is mainly tasked with funding schools and libraries, while the broadband project is mostly involved with raising computer literacy and promoting Internet usage. The broadband map is used to build a broadband monitoring platform based on geographic information systems.

Projects under the USDA include both the so-called "last mile" and "middle mile" examples. The focus is on projects that provide "last mile" Internet connectivity to homes, businesses, and other end-users, primarily providing broadband coverage in rural and remote areas. One example is funding the Rivada Sea Lion company to provide 4G service to under-served areas in Alaska (US$25.3 million in government grants and a further US$6.4 million in leverage). Another example is funding the Big Island Broadband/Aloha Broadband to provide broadband services to the 600 residents and companies on the northern islands of Hawaii (US$100 000 in government grants and another US$87 000 funding).

The Broadband Universal Service Fund – The FCC set up the CAF (Connect America Fund) to provide the average American with affordable broadband services at actual download speeds of at least 4 Mbps. The US$4.5 billion that is currently being used to subsidize telephony services in high-cost regions will subsidize broadband services instead. The FCC plans to increase the Universal Service Fund from US$8 billion to US$15.5 billion over the next 10 years in order to support the construction of broadband infrastructure. In addition, a mobile broadband fund will be set up to provide targeted funding support, so that the 3G wireless network coverage in any state in the United States meets the minimum requirements of universal service.

| The European Union | In 2008, US$1 billion in funding was provided to achieve 100% 2 Mbps high-speed Internet coverage by 2010. This was mainly targeted at the white spots (areas with no Internet services) and grey spots (underserved areas) in suburban and rural areas; |

Since 2003, the EU has provided broadband state aid to its member countries year by year. In 2010, the EU approved broadband state aid projects with a total value of EUR1.8 billion euros, more than four times the amount approved in 2009, as shown in the figure below.

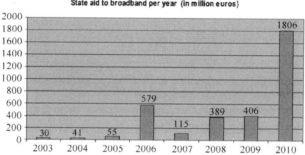

State aid to broadband per year (in million euros)

The European Commission announced on 19th October 2011 that the EU would roll out a large-scale investment program amounting to EUR50 billion. It would be used to develop transportation, energy, and broadband networks to ensure future development and growth in employment. Of the EUR50 billion, EUR9.2 billion would be spent on building high-speed and ultra-high-speed broadband networks in Europe to improve digital services. The program will attract an even larger amount of private and public investment in infrastructural construction and services. The European Commission estimates that every Euro invested will attract another 6 to 15 Euros. This means that the initial broadband funding of EUR9.2 billion is expected to attract 50-100 billion Euros of investment in the construction of broadband networks.

| Japan | Between 2008 and 2009, the Japanese government invested a total of 37.1 billion Yen in intelligent transportation systems, improving ICT network facilities and training, and the construction of broadband infrastructure in rural areas. |

The Japanese government provides preferential tax policies to broadband access operators, including corporate tax redemption, and tax incentives for the depreciation and amortization of fixed assets. It also provides debt guarantees for operators that provide broadband access and low-interest financing through the Development Bank of Japan.

South Korea	In February 2009, the South Korean government announced plans to provide 1 Gbps broadband services by 2012. This would require a total of 34.1 trillion Korean Won over a five-year period, with 1.3 trillion coming from the central government and the rest provided by private operators. The government had previously announced in 2006 that it would invest 26.6 trillion Korean Won in FTTH (Fiber To The Home), fiber-optic LAN (Local Area Network), and FHC. In order to build broadband networks in rural areas, the government provided KT (Korea Telecom) with a low interest loan of US$77 million previously.
Brazil	The Brazilian government has announced that it will invest 11 billion riyals (about US$6.1 billion) in the PNBL, of which 3.22 billion riyals (or US$2.03 billion) was invested in Telebras, the former state-owned telecom operator, in the four years since 2011. Meanwhile, the national development bank, BNDES, will provide 7.5 billion riyals in loans to Telebras and other operators. In addition, Funntel (Brazil's national telecommunications investment fund) will provide 1.75 billion riyals to fund relevant research and development projects to launch a broadband program. State-owned Telebras is the main undertaker of the PNBL program, with private operators playing supplementary roles. On 30th June 2011, Oi, Telesp TLPP4.SA, CTBC and Sercomtel – the four largest private operators in the country – also announced their participation in the PNBL by high-speed Internet access of at least 1 Mbps at a price of 29.80 riyals/month in tax-exempt states and 35 riyals (US$22)/month in non-tax-exempt states.
Finland	In October 2009, the Finnish government formally proposed that broadband access be included among citizen's basic rights. Since 1st July 2010, this proposition became law, making Finland the first country in the world to recognize the right to broadband access through legislation. All network service providers in Finland are obliged to provide users with 1 Mbps broadband Internet access service regardless of their location in the country. According to the Finnish Ministry of Communications, this law is one of the most important achievements the government has made in regional policies. The Finnish government hopes to achieve universal broadband service and attain its goal of 100 Mbps broadband access for 99% of the country through private capital, government subsidies, local government funding, and investment from the EU. Among these, subsidies from the federal government will account for no more than a third, with the local government and the EU providing another third, and finally at least a third from private investment. Altogether, the government has invested about US$131 million. According to the Finnish government, the goal for the first phase was to have broadband speeds of at least 1 Mbps by 2010. To this end, it will provide a third of the funding, with the remaining two thirds coming from local governments, the EU, and telecom operators. The construction of the fiber-optic network in this project will cost about EUR200 million, and the Finnish government will provide EUR67 million.

Australia	In Australia, the government has set up a new company called the National Broadband Network Company (NBN Co), and owns 51% of it. NBN Co is responsible for constructing a nationwide broadband network before leasing broadband services to other companies. Investment in the construction of the Australian National Broadband Network (NBN) included a total capital outlay of AU$35.9 billion, of which AU$27.5 came from the government, with the balance obtained through debt and income. For at least 10 years, the government will own 51% of the shares in NBN Co, and will be responsible for operating the network. There is also a dedicated parliamentary committee overseeing the network construction, which is expected to take eight years. Network operators are also promised equal access to all facilities. According to the Australian government, the cable installation cost of the NBN is around AU$35.9 billion – around AUD$7 billion less than the preliminary estimate of AU$43 billion. To support the construction of the NBN, existing network resources were fully utilized to reduce the amount of investment required. In late November 2010, following intense negotiations and heated debates that lasted for several days, the Australian Senate finally approved a critically important government proposal to allow the separation of Telstra's retail and wholesale operations, and to allow NBN Co to use Telstra's underground infrastructure. Telstra will become an NBN user, paving the way for the NBN project. The then Telstra CEO, David Thodey, stated his willingness to work with the Australian government in support of the spinoff bill. In the next ten years, the government will pay Telstra AU$13.8 billion as compensation for its gradual exit from the market and for the use of its infrastructure and facilities for the construction of the NBN.
Russia	In July 2010, the Government Information Technology Committee of Russia approved a draft of the long-term plan for the implementation of an "Information-based Society from 2011 to 2020" by the Russian Ministry of Communications. Russia plans to invest 10 billion Russian rubles (about RMB2.06 billion) annually for the next 10 years in the project, making a total investment of 100 billion Russian rubles.
United Kingdom	Two thirds of the funds needed for the construction of the broadband network come from private enterprises, and the capital is exempted from tax in the first year. Funding for most rural areas comes from the public purse. The government plans to contribute £830 million, some of which comes from the funds set aside to help pay for the BBC's going digital. £50 million of the £830 million will be used to fund broadband access trials in remote areas such as North Yorkshire, Herefordshire, and the highlands and islands of Cumbria. In November 2011, the British government allocated another £100 million to install super-high-speed broadband in 10 cities.

1.2 National Broadband Development Strategies

Currently, countries around the world are competing to seize the opportunities provided by broadband as the focus of their strategy to gain technological advantages. At present, broadband is driving the robust development of a new generation of information technology industries, and is shaping a new competitive edge in both the manufacturing and service industries. As such, it has become a strategic and key infrastructure into which national technological resources are directed, and in which technological innovations occur. To seize the opportunities offered by the digital economy, countries are giving priority to building their broadband networks.

For example, the United States invested US$7.2 billion in 2009 to support broadband development; the EU invested EUR9.2 billion in 2011 to support the construction of high-speed broadband network infrastructure and public service platforms; in addition to the £530 million made available previously, the United Kingdom allocated another £100 million in 2011 towards the construction of super-high-speed fixed and mobile broadband networks in 10 cities; Brazil invested US$7.3 billion in 2010 to popularize broadband usage amongst low-income families; and in 2010, India invested US$3.09 billion to develop high-speed broadband networks in rural areas.

According to the latest statistics from the International Telecommunication Union (ITU), at the end of 2012, major developed countries and 127 developing countries in the world had proposed and implemented national strategies or action plans for broadband development. Both financial and policy support had also been given, with an emphasis on supporting the deployment of ultra-high-speed networks and increasing penetration in rural areas. The focus is also on supporting the enhancement of broadband applications and business innovation so that broadband can be better used to boost economic growth. In addition to direct government investment, more than 40 countries have already established long-term broadband universal service mechanisms.

1.2.1 The European Union

1) Strategic positioning

At present, there are more than 250 million Internet users in the EU member states, and the broadband utilization rates of member countries such as the United Kingdom, Germany, France, and Sweden are among the highest in the world. Broadband usage rate across the EU is already at 24.8%, topped only by Japan, South Korea, and the United States.

With the acceleration of broadband development around the world, the EU hopes to closely integrate its own with the mobile Internet by building upon its strong leadership in terms of wireless and mobile communications development. In so doing, it hopes to buck the current Internet development trend of following the United States and eventually cement its own leading position.

After the global financial crisis, the EU placed strategic importance on the development of information technology to revive its economy. In March 2010, the European Commission promulgated the Europe 2020 Strategy to establish the European Digital Agenda as one of the seven flagship programs for promoting economic growth in the EU. Its goals were to increase the contributions made by informatization to Europe's economy and society by building on its foundation of high-speed and ultra-high-speed Internet; to achieve universal broadband access by 2013; and for all Internet connections to achieve speeds of more than 30 Mbps by 2020. The European Commission estimates that telecom operators need to invest EUR250 billion (US$318 billion) to achieve these goals.

The EU broadband strategy pays particular attention to the development of wireless broadband networks based on the existing advantages of the EU in the field of mobile communications. Wireless broadband is increasingly important to the EU telecommunications market because it incorporates the best features of broadband and wireless networks. It is also cost effective and has a short construction period, fast service provision, and relative flexibility. It allows the dynamic allocation of system resources, and incurs low system maintenance costs. At the same time, almost every European owns at least one cell phone. Europe is a mature market with a natural user base, and is ripe for the development of wireless broadband.

2) Goals of the EU Broadband Development Strategy

The EU's Europe 2020 Strategy establishes the European Digital Agenda as one of the seven flagship programs for promoting economic growth in the EU. Its goal is to increase the contributions made by of informatization to Europe's economy and society by building on its foundation of high-speed and ultra-high-speed Internet. Specifically, in the Europe 2020 Strategy, the EU's developmental goals for its broadband strategy fall into three phases.

The first phase (by 2013) comprises short-term basic goals such as having at least 14 million FTTH (Fiber to the Home) subscribers in EU member states by the end of 2012, and achieving universal access to broadband throughout the EU by 2013.

The second phase (by 2015) comprises medium-term goals. By 2015, 50% of EU citizens will be able to shop online, and 20% will have access to cross-border online services. By 2015, the Internet utilization rate will increase from 60% to 75% while the rate for those with disabilities will increase from 41% to 60%. By 2015, the number of EU citizens who have never used the Internet before will fall from 30% to 15%. Finally, at least 50% of EU citizens will be able to make use of public services online by 2015.

The third phase (by 2020) comprises the ultimate goals. All internet connection speeds will exceed 30 Mbps by 2020, with at least half of all European households having access to broadband speeds of up to 100 Mbps. By 2020, the total annual investment by all EU member states on ICT research and development will reach EUR11 billion.

3) Measures taken by the EU to develop broadband strategy

To achieve the goals set out in the Europe 2020 Strategy, the EU hopes to work jointly with its member states and also coordinate cooperation among them. Specific measures such as encouraging and increasing investment, developing wireless broadband, and making mindful use of the development funds will promote the pan-European development of broadband communications, thereby promoting the informatization of the entire EU.

(1) Reducing broadband investment costs

Capital investment is the primary driver of broadband development. The EU actively encourages its member states to increase investment in the development of broadband communications at various levels, including national and regional. At the same time, the EU recommends that its member states actively seek ways to reduce the costs of broadband construction and development. It has made the following recommendations to its member states and the relevant departments at all levels:

- Reduce investment costs by improving information transparency, reducing barriers to information, and deploying the relevant resources mindfully, making effective use of existing resources, and preventing duplication.
- Reduce the investment cost of broadband construction by eliminating administrative obstacles such as layers of approval for access rights to infrastructural projects like new base stations, and the existing difficulties in renewing contracts.

(2) Promoting the development of wireless broadband

Leveraging its years of experience in wireless broadband development, the EU hopes to promote the strategic development of the entire field through its focus on wireless broadband. To this end, spectrum is an important resource. The EU is trying to establish a ubiquitous pan-European network of wireless broadband and cable broadband through the reasonable allocation of spectrum. Regarding the allocation of wireless broadband spectrum resources, recommendations were as follows:

- The European Commission recommends that EU member states provide mobile operators with a portion of the valuable broadcast frequencies used by television stations before 2013, in support of the creation of an EU-wide wireless broadband service market. This proposal is part of the EU broadband network reform plan requiring its 27 member states to allocate the 800GHz band to the mobile broadband network by January 2013.
- With the acceleration of global broadband development, the EU must increase its flexibility and competitiveness in the mindful allocation of spectrum resources. Encouraging rapid utilization and allowing secondary transactions can maximize the use of this scarce resource. The pan-European wireless broadband network and wired broadband will promote the informatization of society, and is also a mean of boosting the economy and enhancing the EU's core competitiveness.

(3) Mindful use of the broadband development fund

The EU supports the construction and development of broadband communications within its borders through the establishment of the SRD (Structural and Rural Development) fund. From 2007 to 2013, a total of EUR2.3 billion in SRD funds are slated to be invested in the construction and development of broadband. The EU has given the following recommendations for the management, allocation, and promotion of the SRD funds.

- In 2011, a guide to broadband investment was released to encourage and guide member states and their relevant departments to apply for and effectively use broadband development funds and investments.
- In 2011, parties from both inside and outside of the industry were invited to participate in broadband development projects supported by the SRD. Their opinions on broadband development were sought.
- The European broadband portal was restarted and expanded to provide a multilingual broadband platform. This platform makes it convenient to exchange information related to broadband development projects, and provides guidance on issues such as rules for state aid and the implementation of the regulatory framework.

1.2.2 The United States

1) Strategic Positioning

Making use of its first-mover advantage in the Internet field, the United States has been the global leader in information and communications since the 1980s. At this critical juncture, during a period of global transformation in information and communication technologies as well as industrial integration and transformation, the United States hopes to maintain its leading position in the information-based society of the future through its broadband development strategy. It thereby hopes to promote both technological and business innovation in information and communication, and to construct a new national information infrastructure. This is especially vital at this crucial point in tackling the global financial crisis.

The United States has a solid foundation for developing its broadband strategy. According to a report released by the US Department of Commerce in February 2010, broadband penetration in the United States reached 63.2% in 2009 – a significant increase of 13% over 2007. However, broadband quality in the United States is relatively poor, and appears to be improving relatively slowly. According to a report by the Communications Workers of America (CWA), the broadband access rate in the US is 5.1 Mbps, up by only 1.6 Mbps from 3.5 Mbps in 2007.

Additionally, broadband users in the US are highly differentiated in terms of income, age, race, and geography. At 29.9%, the broadband penetration of households with an annual income of less than US$15,000 is the lowest, while that of households with an annual income of more than US$150,000 is the highest, at 88.7%. Broadband penetration among households in the 18-24 age bracket is the highest at 80.8%, while that of households in the over-55 age bracket is the lowest at 46%. Meanwhile, the broadband network infrastructure in many suburbs and rural areas of the United States is far from perfect. As of the end of 2009, the broadband penetration of rural households was 54%, 12% lower than in urban areas.

2) Strategic Goals

In February 2009, Obama signed the American Recovery and Reinvestment Act (ARRA 2009) into law in response to the global financial crisis, and US$ 7.2 billion was designated for broadband development. In April 2009, the FCC (Federal Communications Commission) began working on a strategic plan for broadband development in the United States. On 15th March 2010, the FCC submitted its National Broadband Plan to Congress, including input from US citizens. Six goals were set out in the plan. The focus of the latest strategic broadband development is to improve infrastructure, increase speed, expand coverage, and ultimately realize universal access. These goals are as follows:

- At least 100 million US households should have access to affordable broadband plans by 2010, with actual download speeds of at least 100 Mbps and actual upload speeds of at least 50 Mbps.
- The United States should lead the world in mobile innovation, and have the fastest and most extensive wireless network worldwide. Every American should be able to afford access to broadband services.
- Every American should have access to excellent broadband services, and possess the technical knowledge necessary to select the most appropriate.
- Every community in the US should have broadband services of at least 1 Gbps in order to support institutions such as schools, hospitals, and government agencies.
- To ensure the safety of the American people, every first aider should have access to wireless networks nationwide. These public safety broadband networks should be interoperable.
- To ensure that the United States leads the clean energy economy, every American should be able to use broadband to track and manage their energy consumption in real time.

3) Strategic Initiatives

The US government plans to concentrate on four aspects to achieve these six long-term goals, with a view to ensuring the healthy development of the broadband ecosystem.

- The creation of a mechanism that maximizes the interests of consumers through healthy competition, while promoting innovation and investment. This involves:

  · Collecting, analyzing, and publishing details of broadband service prices and competition in each market;
  · Requiring broadband service providers to release information on the price and delivery standards of their broadband services so that consumers can choose the best service on the market;
  · Conducting a comprehensive evaluation of the competitive regulations;
  · Opening up and allocating additional spectrum resources for license-free usage;
  · Increasing the capacity of broadband services in urban zones and coverage in rural areas;

- Taking action to determine how best to achieve extensive and seamless broadband coverage that is also competitive;
- Improving the relevant laws and regulations to create a competitive and innovative video set-top box market;
- Fully protecting consumer privacy.

- Developing broadband infrastructure and reducing barriers to competition through the effective allocation and management of state-owned assets. This involves:

  - Recovering the 500 MHz spectrum within 10 years, and allocating the 300MHz spectrum for mobile network usage within 5 years;
  - Encouraging spectrum auctioning;
  - Ensuring that spectrum allocation and usage become more transparent;
  - Intensifying research into new spectrum technologies;
  - Promoting the use of broadband infrastructure in the United States by improving the management of access rights;
  - Implementing policies such as "dig-once" for the development of effective infrastructure;
  - Providing the Ministry of Defense with ultra-high-speed broadband to develop next-generation broadband network applications for the military.

- Setting up the Connect America Fund to popularize broadband (with actual download speeds of at least 4 Mbps) and voice services that are affordable for the general public. This involves:

  - Establishing the Connect America Fund to ensure universal access to broadband networks by the general public;
  - Creating a mobile network fund to ensure that every state can achieve the average standards of 3G wireless network coverage;
  - Improving the inter-carrier (spectrum) compensation system;
  - Designing the new Connect America Fund to reduce taxes so as to narrow the broadband gap and ease the burden on consumers;
  - Establishing a mechanism to ensure that broadband services are affordable for low-income families.

- Improving laws, policies, standards, and incentives to maximize the benefits of broadband in key government departments. This includes:

- Healthcare – Improving the quality and reducing the cost of healthcare through broadband services;
- Education – Enabling students to carry out distance learning and access online content through the use of broadband services, to improve public education;
- Energy and environment – Taking advantage of innovations in broadband technology to reduce carbon emissions and improve energy efficiency, thereby reducing the US's dependence on oil imports;
- Economic opportunities – Improving access to jobs and training, thus supporting the development of enterprises;
- Government operations and citizen participation – Making government services and internal processes more effective, and improving the quantity and quality of citizen engagement;
- Public and national security – Providing first aiders with timely access to relevant information.

1.2.3 Japan

Japan's broadband access market has enjoyed rapid growth since 2000. It currently leads the world in all aspects, including broadband penetration, access speed, and variety of applications. When it comes to the rapid development of broadband in Japan, the continued support of the national strategy is a very important driver.

Prior to 2009, the Japanese government released the e-Japan Plan (2001–2005), the u-Japan plan (2004–2010), and new strategies in IT development. In July 2009, Japan launched its national informatization strategy, known as i-Japan 2015, with an investment of USD$1.9 billion. By 2015, it will achieve fast and simple network access at fiber-optic (Gb) speeds, and high quality ultra-high-speed broadband infrastructure that is very stable will be built. In September 2010, the Japanese government announced in its Blueprint for a New Economic Development Strategy that it intended to make broadband network services available to the approximately 49 million households in the country by 2015 through developing infrastructure.

1.2.4 South Korea

As the country with the highest broadband penetration in the world, South Korea sees the importance of diversified development and application of broadband networks, and closely integrates it with the development of IP telephony, 3G/4G wireless Internet access, and digital terrestrial television broadcasting.

According to the Medium to Long-Term Plan for the Development of the Broadcasting and Communications Network for 2009-2013 promulgated by the South Korean government in January 2009, the government will invest US$32.5

billion to convert 60% of the existing phone lines to IP telephony by 2013 and popularize VoIP (Voice over Internet Protocol). By 2012, South Korea's 14 million Internet users will enjoy wired Internet service between 50 Mbps and 100 Mbps. After 2012, an ultra-high-speed broadband network will be built to provide 1 Gbps wired Internet access. 1 Mbps 3G wireless access will be provided to 40 million users, followed by 10 Mbps 3.9/4G services by 2013. In addition to building IPTV by 2010, the infrastructure for terrestrial television broadcasting will be enhanced to a two-way interactive environment that allows users to shop online while watching TV.

According to a report from the Communications Workers of America, from the perspective of the broadband network access mode, the number of ultra-broadband users in South Korea exceeds 16 million. At present, 95% of all South Korean households have broadband access with an average access rate of 20.4 Mbps, placing the country top in the ranking of global broadband performance. However, the South Korean government is not resting on its laurels. Instead, the country plans to build a gigabit broadband network that would allow a DVD-quality movie to be downloaded in 10 seconds.

1.3 The Current State of Broadband Development in China

In recent years, significant progress has been made in the development of China's broadband, making important contributions to the national economy and social development.

1) The Increasing Role of Broadband in the National Economy and Social Development

• Broadband as an emerging force in promoting economic growth.

In 2011, direct investments in fixed broadband and 3G networks exceeded RMB220 billion. Expenditure in information services was nearly RMB500 billion, driving related upstream and downstream industries to realize an output value in excess of RMB2.4 trillion.

• Broadband as an important foundation for promoting the transformation and development of traditional industries, as well as fostering emerging industries.

With e-commerce transactions totaling nearly RMB6 trillion, and emerging industries such as software outsourcing, information services, cloud computing, and the Internet of Things booming, the optimization and upgrading of China's economic structure is set to receive a substantial boost.

• Broadband as an important channel for increasing employment rates.

In 2011, the development of fixed broadband and 3G networks led to more than 1.7 million job openings in areas such as communications equipment, construction, and service development. It accounted for nearly 10 million jobs being created in related industries such as e-commerce and logistics.

• Broadband as an important means of support for the improvement of public services and social management.

Educational training, health care, e-government, and social security based on broadband networks are widely used.

2) Significant Improvement to the Industrial Chain of the Broadband Network

China has developed industrial support capabilities in a variety of fields, including optical communications, broadband wireless communications, next-generation Internet, mobile internet, and cable TV. It covers many aspects such as systems, terminals, chips, key devices, and instrumentation. The country also possesses mainstream technologies such as fiber-optic broadband access, high-capacity long-distance transmission, high-end routing switch, and 3G mobile communications. China's ability to integrate and innovate in the area of broadband technology has also improved significantly, with some domestic broadband equipment attaining advanced global standards. In new areas such as TD-LTE and IPv6 it has also achieved breakthroughs in core technologies. Lastly, its role in setting international standards continues to grow.

3) Sustained Growth in Broadband Network Coverage and Penetration

• Significant increase in network capacity.
  Broadband is available in all cities and townships, and in 84% of the administrative villages across the country. More than 90% of broadband subscribers enjoy access rates above 2 Mbps, and nearly 40% of cable TV connections are bidirectional. International bandwidth reached 1.4 Tbps in 2011, which is 10 times that of the initial period of the 11th Five-Year Plan.

• Significant growth in penetration.
  In April 2012, the number of fixed broadband users reached 159 million – an increase of 3.2 times that of the initial period of the 11th Five-year Plan, while broadband penetration for households increased to 36.7%. The number of 3G users reached 159 million, with the proportion among mobile subscribers rising to 15.4%.

- Expanding the scale of usage.

    In April 2012, the number of netizens stood at 532 million – an increase of 4.8 times that of the initial period of the 11th Five-Year Plan. Penetration rose to 39.7%, while the number of web pages reached 86.6 billion – an increase of nearly 30 times.

4) Continued Strengthening of Information Security and Network Capabilities

- Security for network information is becoming more robust.
- Security technology, backup, and emergency response capabilities are consistently being enhanced.
- Strengthening of fundamentals such as security ratings and protection, safety evaluation, and risk assessment of basic networks and important information systems.
- Significant improvements to the controllability of key network equipment and the competitiveness of the network information security industry.

5) Keeping up with Other Countries

- China has not kept up with the global pace of broadband development. The gap between the penetration of broadband networks in China and in developed countries increased from 10% in 2005 to 12.8%, and access speed is below the global average. At present, China's broadband penetration is only half that of developed countries (25.6%); for most people, the Internet access rate is less than a quarter of that in developed countries. Data from the International Telecommunication Union shows that in the access rate ranking of the ICT Development Index (IDI), China's rank dropped from 64th in 2007 to 82nd (down by 18 places) while India's rank rose from 129th to 116th (up by 13 places), and Brazil's rose by 3 places from 69 to 66. The gap between China's broadband development and that of the rest of the world is widening.
- In both rural areas and remote areas, the digital divide has continued to expand due to poor natural conditions, a scattered population, low economic levels, and high network deployment costs. It is quite challenging to build, operate, and maintain networks relying solely on market mechanisms. By the end of 2012, the broadband penetration in the central and western regions will lag behind that of the eastern regions by 6%, while the rural broadband penetration will only be 5–12% lower than the urban broadband penetration.
- There is still a large gap between the expected and actual broadband network performance. Therefore, there is an urgent need to improve the access bandwidth of websites, the layout and capacity of content distribution networks, and the broadband access rate of subscribers, to improve the online experience.

- There are not enough broadband application services. There is still an urgent need for the integration of information technology and industrialization, the development of small and medium-sized enterprises, rural development, and the informatization of social and public services.
- Broadband development places new demands on high-speed, ultra-large capacity and green energy saving for the next-generation broadband technology industry. High-end optical devices and key chips in the upstream of the broadband optical communication industrial chain are greatly dependent on foreign countries, and this restricts further development. Hence, there is an urgent need for breakthroughs in key technologies as well as enhanced independent innovation and industrial support.
- It is difficult to upgrade the broadband networks in old residential districts in urban areas owing to a lack of policies covering right of way, site-selection of base stations, and connecting households to the network.

The underlying reasons are as follows:

- Compared with other countries, China's understanding of the urgency for broadband development is limited, the status of broadband infrastructure is still not clear, and insufficient social resources have been allocated.
- Broadband development lacks top-level design and overall planning.
- The competitive environment in the broadband market needs to be improved, and institutional mechanisms need to be created.

1.4 China's Broadband Development Goals

According to the "Broadband China" Strategy and Implementation Plan (Issue No. 31 of 2013) issued by the State Council in August 2013, the objectives of broadband development in China are as follows.

The first phase of the next-generation national information infrastructure – designed to meet the needs of economic and social development – was scheduled to be completed by 2015. The aim of having fiber-optics in urban areas and broadband in rural areas would basically be realized. Fifty percent of households would have fixed broadband, 32.5% of mobile users would be using third-generation mobile communications (3G / LTE), 95% of administrative villages would have access to broadband (wired or wireless, as below), and practically all non-profit organizations such as schools, libraries, and hospitals would have broadband access. Access rates for urban and rural households would reach 20 megabits per second (Mbps) and 4 Mbps respectively, with the access rate going up to 100 Mbps in some developed cities. Broadband usage levels would increase significantly, and mobile Internet would be prevalent. At the same time, network and information security capabilities would be significantly enhanced.

By 2020, the gap between the development levels of China's broadband network infrastructure and that of developed countries will be substantially narrowed. Citizens will enjoy the benefits of economic growth, convenience, and development opportunities brought about by broadband. Networks will be available in all urban and rural areas, with 70% of households having fixed broadband, 85% of mobile users having 3G/LTE, and more than 98% of administrative villages having access to broadband. Access rates for urban and rural households will reach 50 Mbps and 12 Mbps respectively, with some households in developed cities enjoying an access rate of up to 1 gigabits per second (Gbps). Broadband usage will be a part of everyday life, and mobile Internet will be ubiquitous. Technological innovation and the competitiveness of the broadband industry will reach advanced global standards, forming a relatively robust network and information security system.

The technological roadmap and timeline of China's broadband development are as follows. In sync with the evolution of broadband technology, it will:

- make full use of the existing networks;
- focus on overall economic and social development needs and broadband development goals;
- strengthen and improve the overall layout;
- systematically resolve key issues such as access rates, coverage, and usage levels;
- strengthen industrial development and security;
- continue to improve the overall level of broadband development;
- comprehensively enhance broadband's ability to support sustainable economic and social development.

1) Technological Roadmap

China will carry out systematic development of broadband networks in phases through:

- coordinating the construction of access networks, metropolitan area networks (MAN), and backbone networks;
- comprehensive utilization of both wired and wireless technology;
- commercial deployment of the next-generation Internet based on the Internet Protocol Version 6 (IPv6).

It will build networks with high-speed access, extensive coverage, and smooth adaptation to local conditions. It will also make use of technologies such as fiber-to-the-home and fiber-to-the-building to construct and improve access networks in urban areas. These access networks can then be combined with 3G/LTE and wireless LAN technology to achieve seamless broadband network coverage. As for rural areas, technologies such as wired or wireless broadband are chosen subject to the local conditions for the construction of access networks.

China will also construct MANs using high-speed transmission, integrated load-bearing, intelligent sensing, and security control. It will promote the use of high-speed transmission, packet transmission, and high-capacity routing switching technologies in MANs. It will also expand the bandwidth of MANs and increase the amount of traffic they can handle. Meanwhile, it will promote intelligent network transformation, and raise the multi-service load-bearing, sensing, and safety control standards of MANs.

It will also build backbone networks through optimizing structures, enhancing capacity, implementing intelligent scheduling, and insisting on efficiency and reliability. Other tasks include optimizing the backbone network structure, improving the international network layout, and promoting ultra-high-speed wavelength-division multiplexing systems and cluster router technologies. China will also enhance the backbone network's capacity and intelligent-scheduling capability, and ensure the high-speed, high-efficiency, safe, and reliable operation of the network.

2) Timeline of Development

(1) Acceleration Phase (until the end of 2013)

The focus is on enhancing the construction of fiber-optic and 3G networks, improving the broadband network access rate, and improving and enhancing the online experience. To this end, China will:

- proceed with the upgrade to fiber-optics in urban areas;
- choose between wired and wireless broadband in rural areas to speed up the construction of broadband access networks in administrative villages, raise access speeds, and provide cost-effective broadband access;
- improve the quality of 3G networks in urban areas;
- expand the coverage of 3G networks in rural areas;
- expand the scale of testing in time-division duplex mobile communication for long-term evolution (TD-LTE);
- speed up the construction of next-generation broadcasting networks;
- replace copper wires with fiber-optics;

- move towards bi-directional networks to promote interconnection and interoperability;
- simultaneously expand and upgrade metropolitan networks;
- optimize the backbone network of the Internet by focusing on interconnectivity;
- upgrade and improve websites to raise access rates.

By the end of 2013, there would be more than 210 million fixed broadband users. The fixed broadband penetration of urban and rural households would reach 55% and 20% respectively. There would be more than 330 million 3G/LTE users with 25% user penetration. Ninety percent of administrative villages would have broadband access. Eighty percent of broadband users in urban areas would enjoy broadband access rates of up to 20 Mbps, while 85% of broadband users in rural areas would enjoy broadband access rates of up to 4 Mbps. Wireless broadband network coverage in both urban and rural areas would be significantly improved. Wireless local area networks for hotspots would be available in important public places in urban areas. Up to 60% of all users of the national cable television network would enjoy interoperable and interconnected platforms.

(2) Promotion Phase (2014–2015)
The focus is now on accelerating the expansion of coverage and the scale of broadband networks, and increasing usage levels. At the same time, access rates will continue to increase. The aims are to:

- speed up expansion both in terms of coverage and scale of fiber-to-the-home networks in urban areas;
- actively make use of wireless technologies to speed up the rolling out of broadband networks in administrative villages in rural areas;
- promote fiber-to-the-village where possible;
- continue to widen both the coverage and depth of 3G networks;
- promote the widespread use of TD-LTE in business;
- continue to promote the construction of the next-generation of broadcasting networks;
- further expand the coverage of such networks and speed up the development of interconnection and interoperability;
- fully optimize the national backbone network;
- enhance research and development of new technologies in key areas such as optical communications, wireless broadband communications, next-generation

Internet, next-generation broadcasting networks, and cloud computing, and produce innovative outcomes in some of these key areas.

By 2015, there would be more than 270 million fixed broadband users, while the fixed broadband penetration of urban and rural households would reach 65% and 30% respectively. There would be more than 450 million 3G/LTE users with 32.5% user penetration. Ninety-five percent of administrative village would have broadband access. Broadband users in urban areas would enjoy broadband access rates of up to 20 Mbps, while those in rural areas would enjoy broadband access rates of up to 4 Mbps.

The 3G network would cover both urban and rural areas, while LTE would become widespread in business. Wireless local area networks to provide coverage for public hotspots would be fully realized, with a comprehensive upgrading of service quality. The number of Internet users in the country would reach 850 million. Their ability to access the Internet and enjoy high-quality service standards would rise significantly. Up to 80% of all users of the national cable television network would enjoy interoperable and interconnected platforms. The quality of the interconnection between backbone networks would meet business development needs, as would the broadband access bandwidth and service quality offered by Internet service providers. Crucial core technologies with intellectual property rights would be acquired in key areas such as wireless broadband communications and cloud computing.

Broadband technology standards would be refined over time, gaining a significantly greater influence on the setting of international standards.

(3) Optimization Phase (2016-2020)
In this phase, the focus will be on promoting the optimization of broadband networks, as well as the evolution and upgrading of broadband technologies. The aim is to reach world-class standards in service quality and applications, as well as improving the overall capability of the broadband industry.

By 2020, the infrastructure for convenient, high-speed, and advanced broadband networks covering both urban and rural areas will mostly be completed. There will be more than 400 million fixed broadband users, and household penetration will reach 70%, with fiber-optic networks for all households in urban areas. There will be more than 1.2 billion 3G/LTE users with 85% user penetration. Ninety-eight percent of administrative villages will have broadband access, and this will be extended to villages using a variety of technical means if possible.

Broadband users in urban and rural areas will enjoy broadband access rates of up to 50 Mbps and 12 Mbps respectively. Fifty percent of households in urban areas will enjoy access rates of up to 100 Mbps, while some households in developed cities will enjoy access rates of up to 1 Gbps. LTE networks will cover most urban and rural areas. The number of Internet users in the country will reach 1.1 billion, while their ability to make use of the Internet, and the service standards they enjoy will rise significantly.

Up to 95% of all users of the national cable television network will enjoy interoperable and interconnected platforms. Comprehensive breakthroughs will be achieved to overcome the bottlenecks within the high-end infrastructure industry that were restricting broadband development. Research and development in broadband technologies will reach advanced global standards, and a broadband industrial chain will be built with a sound structure and international competitiveness, comprising a group of world-leading innovative enterprises.

1.5 Developing the Next-generation Internet Within the "Broadband China" Strategy

According to the technology roadmap of the 'Broadband China' Strategy and Implementation Plan (Issue No. 31, 2013), China will carry out the systematic development of broadband networks in phases. This will involve coordinating the construction of access networks, metropolitan area networks (MAN), and backbone networks; the comprehensive utilization of both wired and wireless technology; and the commercial deployment requirements of the next-generation Internet based on the Internet Protocol Version 6 (IPv6).

Within the promotion phase (2014–2015) of the "Broadband China" strategy, the important strategic tasks are technological innovation and large-scale deployment in fields such as the next-generation of Internet. The plan is to fully optimize the national backbone network, and enhance research and development of new technologies in key areas such as optical communications, wireless broadband communications, the next-generation Internet, next-generation broadcasting networks and cloud computing. The wider aim is to produce innovative outcomes in some of these key areas.

Accelerating the development of the next-generation of Internet is vital to the implementation of the "Broadband China" strategy.

During this critical period, in which major changes have taken place worldwide in information and communication technologies, and in the face of the tremendous shifts that have occurred both inside and outside of China since the onset of the global financial crisis, speeding up the development of the next-generation Internet is of great strategic importance.

1) Leading the development of the next-generation Internet – a major opportunity to seize the global vantage point in terms of the economy and science and technology

Internet development has become an important indicator of a country's overall strength. In response to the profound changes that have taken place in economies and societies worldwide, many countries – especially the major powers – have made Internet development a priority when formulating their own strategic economic plans.

The United States has devised the Ipv6 transition plan to maintain its dominance in the field of Internet technology. In developing the next-generation Internet, the EU, Japan, and South Korea are seeking to seize the wider advantages and opportunities associated with the growth of an information-driven society. Meanwhile, countries like the United States have noted the benefits that other nations and regions – particularly China – could reap from the early deployment of IPv6. As such, the United States views China's development of the next-generation Internet as a bid to occupy a major segment of the market, and is therefore a serious threat to its own dominance. Thus, the next-generation Internet has become an international economic and technological vantage point.

There is an urgent need for China to make progress in developing the next-generation Internet. The number of Internet users in China ranks first in the world, but penetration is only 31.8% – much lower than that in developed countries, which exceeds 70% (74.1% in the United States, 77.3% in Korea, and 74.1% in Japan).

In terms of raising network penetration, there is plenty of space for Internet development in China. As the assignment of IPv4 addresses approaches its limit around the world, the circumstances are even more dire in China; all of the current IPv4 addresses will be taken up within one to two years. The development of IPv6 networks is imminent. The practical issue of this shortage of addresses has forced China to the front in the development of next-generation Internet. This will enable it to take the lead in realizing the large-scale commercialization of the next-generation Internet around the world. It also provides a rare and historic opportunity for China to hold the vantage point in terms of international economic and technological strategy.

Since the construction costs of IPv6 and IPv4 networks are basically the same, the sooner China develops the next-generation Internet featuring IPv6, the more conducive it will be to the construction and development of its national infrastructure. This will also reduce the cost of technological upgrades, and will boost China's strength in the market.

2) Developing the next-generation Internet – an important precursor to fostering emerging strategic industries

Speeding up the cultivation and development of emerging strategic industries in response to current trends in the global economy and the new technological revolution, which continues unabated, is an important strategic move by the Central Committee of the Communist Party of China and the State Council to seize the vantage point in this new round of development. It is also a crucial measure to enhance China's long-term competitiveness.

The information industry is one of the most important emerging strategic industries. China has chosen a new road to industrialization – one which integrates information technology and industrialization. As such, there is a need to tackle the dilemma of an information industry that is huge but weak. To this end, there is an urgent need to locate a breakthrough point and find new areas of growth in order to achieve leapfrog development. The development of IPv6 as the core of the next-generation Internet provides an opportunity to do this. Therefore, the next-generation Internet is an important precursor to the development of emerging strategic industries. It can trigger new societal demands and prompt the information industry to adjust its structure and transform its mode of development. It can also drive and nurture other emerging industries, promote industrial technological innovation, and give China the early-mover advantage in future development. At the same time, the development of the next-generation Internet also provides a platform to support the development of other strategic industries.

Developing the next-generation Internet can also further promote the development of emerging industries such as the Internet of things (including sensor networks) and mobile Internet. It can boost China's competitiveness in core technologies such as sensors and RFID chips, narrowing the gap with developed countries. The development of the next-generation Internet will promote the construction of national information infrastructure and promote investment in social informatization, thus creating space for the Internet-equipment manufacturing industry, the software industry, and the information service industry to develop. It will also drive the overall development of the information industry. It is expected that by the end of the 12th Five-Year Plan period, the market size of the next-generation Internet service industry and the software and equipment manufacturing industry will exceed RMB$1 trillion, and will drive the rapid development of other industries.

The next-generation Internet is conducive to tri-network integration. Broadcast networks do not have the burden of being compatible with the existing IPv4 networks, and thus can use IPv6 directly. The transformation into interactive broadcast networks will allow users to move from watching television to using television. The 420 million TV sets nationwide will greatly increase the number of terminals and usage levels for the next-generation Internet in China, elevating the information technology and information industry to a new level.

3) Developing the next-generation Internet – an important impetus for transformation and restructuring

At the 17th National Congress of the Communist Party of China, it was emphasized that the key to achieving economic development goals in future is to make major strides in areas such as accelerating the transformation of the mode of economic development. This will promote the transformation of economic growth from relying mainly on increasing material consumption to depending on scientific and technological progress. Internet technology is a key generic and strategic form of high technology, and lies at the heart of an information-driven society. Prompt large-scale commercial deployment of the next-generation Internet will offer China new opportunities for innovation, and will allow it to seize the vantage point in the core technologies of the next-generation Internet even earlier. It will thereby greatly enhance China's innovation capabilities, accelerate the transformation of its economic mode of development, and further promote industrial restructuring by making use of high-technology to transform traditional industries. Developing the next-generation Internet will speed up the construction of network infrastructure, promote the R&D and industrialization of equipment, and stimulate new business applications. This will bring about significant social investment and consumption demands, and provide new development opportunities for the modern service industry. It will also lay the foundation for the restructuring of the communications industry, so that China can embark on a path of sound economic development that is driven by innovation and endogenous growth.

4) Developing the next-generation Internet – an important way to achieve sustainable development, save energy, and reduce emissions

As more than half of the world's population embraces the modern lifestyle, both energy demands and pressure on the ecological environment have greatly increased. The conflict between fast economic and social development and the limited carrying capacity of the Earth has become increasingly intense. A populous country with rapid industrialization and urbanization, the conflict between energy and ecological environment is particularly prominent in China, which also means that the task of promoting sustainable development is especially arduous. The serious challenges posed to global development mean that there is an urgent need for a new mode of economic development. Countries around the world are actively pursuing green, intelligent, and sustainable development.

Developing the next-generation Internet, especially when it comes to promoting new applications that require a large number of IP addresses such as the Internet of things and the mobile Internet, will help to realize energy-efficiency, eco-friendly lighting, smart power grids, intelligent transportation, and environmental monitoring, thus providing an important path for China to achieve sustainable development, save energy, and reduce emissions. It will also have a profound impact on building a resource-saving and environment-friendly society, coping with global climate change, and safeguarding China's long-term interests.

5) Developing the next-generation Internet – an important foundation for narrowing the digital gap and promoting harmonious development

Due to the unbalanced development of urban and rural areas in China, there is a huge 'digital gap' between the cities and the countryside, and also between the eastern and western regions. This has become an important factor in the development of a smooth-running society.

Developing the next-generation Internet will raise Internet penetration significantly, greatly reducing the digital gap between urban and rural residents, and thereby providing an important foundation for socially cohesive development. It will also contribute to fairness in education and basic healthcare services for all, and improve the level of public services and emergency response capabilities. It will help to build a socialized and networked service system, and raise the efficiency of innovation in society at large, thereby promoting the dissemination, conversion, and application of knowledge. Developing the next-generation Internet will promote all-round social progress and improve the overall quality of life.

6) Developing the next-generation Internet – an important opportunity to deal with cyber hegemony and safeguard China's national security

At present, national security has expanded the dimensions of the sea, land, and air into cyberspace, and network security has become a major security challenge. Internationally, developed countries headed by the United States attach great importance to the development of the Internet, and regard it as an important means to safeguard hegemony. As such, they have established attack and defense systems for network security. Domestically, as an important infrastructure for both the national economy and social development, serious challenges are being posed to Internet security, as well as to national security, which counts the Internet as an important foundation. Therefore, network security has to be an important part of China's national security. Establishing a strong network and information security system is of strategic importance to China's economic growth, social stability, and overall development.

The next-generation Internet provides a new technological platform for solving security problems, and provides development opportunities for the building of an Internet that is more secure and reliable overall. It will also allow China eventually to escape the unfavorable situation of its Internet being constrained by others, and help it to achieve autonomy and control of its network security.

Chapter 2

Development and Evolution of the next-generation Internet - Goals and Pathways

Chapter Highlights:
- *The needs and challenges of the Internet*
- *Developmental goals for the next-generation Internet*

Overview

After more than 40 years of development, the Internet has become an important element of infrastructure for society, and a significant strategic resource for nations around the world. As the cornerstone of cyberspace, the Internet is at a crucial stage of technological transformation. It is evolving into a new generation in order to meet the basic requirements of an information-driven society for safe, credible, controllable, ubiquitous, and reliable cyber infrastructure. The United States and Europe have stepped up their strategic planning, and have strengthened their research and experimentation on future network technologies in order to seize the lead in cyber infrastructure. Their strategic intentions and plans are becoming increasingly clear. As for China, it must seize the rare opportunity brought about by changes in network technology to adjust the technological and industrial pattern to match the dominant power of cyberspace, paying attention the strategic development of cyber infrastructure, and defining its strategic positioning and priorities as soon as possible. In addition, China should speed up its innovation and experimentation in future network technologies, and must enhance its network technology and industrial capabilities in cyber infrastructure so as to shape new competitive advantages against other nations in the information-driven society of the future.

The Internet is in a critical period of technological change and evolution. In its current state, it has several outstanding issues that are recognized by the industry, such as insufficient addresses, inconsistent service quality, lack of a secure and credible mechanism, and poor network controllability. Since the 1990s, both academic and industrial circles have made unremitting efforts to resolve these problems. However, as these problems are determined by the TCP/IP protocol suite adopted by the Internet, they have not been completely rectified. In fact, the rapid popularization and further application of the Internet has only served to highlight the problems. The information-driven society of the future requires cyber infrastructure to be safe, credible, controllable, ubiquitous, and reliable. In the face of these basic requirements, the Internet is facing technological changes, and is in the crucial period of its evolution into the next generation.

At present, the short-term evolutionary pathway of the Internet is relatively well-defined, and the understanding of this matter both in China and abroad is relatively uniform. However, the development pathway for the medium to long term is yet to be clearly defined, and the direction of technological development is still in the exploratory stage.

In the short term, due to the depletion of IPv4 addresses, the development of the mobile Internet and the Internet of things faces a bottleneck, i.e. insufficient addresses. IPv6 can provide an enormous number of addresses, and is currently the only next-generation Internet solution that is mature enough to be used. As such, the industry's consensus is that IPv6 is the starting point for the evolution of the next-generation Internet. In recent years, the development of IPv6 has been accelerating internationally. China has also formulated a roadmap and timeline for the commercial deployment of IPv6, and has embarked on the construction of IPv6 networks. According to the national plan, large-scale implementation of IPv6 will be achieved within the next five years.

As for the medium- to long-term developmental pathway for the Internet, the large-scale commercial deployment of IPv6 will result in the formation of massive networks with large user bases within the next few years. Setting aside this infrastructure to build a new network from scratch is unrealistic. Therefore, the general view is that in the medium to long term, the Internet will evolve from IPv6 networks or networks that are compatible with IPv6. As IPv6 only solves the problem of insufficient Internet addresses and no other problems associated with the Internet in its current form, it cannot promote the next-generation Internet on its own. As such, it is necessary to carry out continuous technological innovations based on IPv6 networks.

The United States and Europe attach great importance to innovation in network technology, and have stepped up their research efforts and support. They are also carrying out research and experimental work on the Future Internet. In China, the Future Internet is generally known as the Future Network or the next-generation Internet. Even though three different terms are used, they are essentially the same, with identical development goals. They are visions for the long-term evolution of the Internet, and the core components of cyber infrastructure.

2.1 The needs and challenges of the Internet

2.1.1 Serious challenges to the sustainable development of the Internet

The Internet was originally designed to satisfy a sole need, i.e. data communication. It was designed to achieve robustness of networks and to support underlying network technologies that were dissimilar. The working assumption was that interconnected users belonged to mutually-trusted communities. Therefore, the traditional architecture of the Internet only supported service to be the best within its capacities. Its design principle was a simple core with an intelligent periphery. The network was deployed intelligently, with terminals at its edge. This type of architecture is simple, but ensures efficient interoperability and good evolution. It is still in use to this day.

Figure 2-1 shows the development of the Internet and its applications.

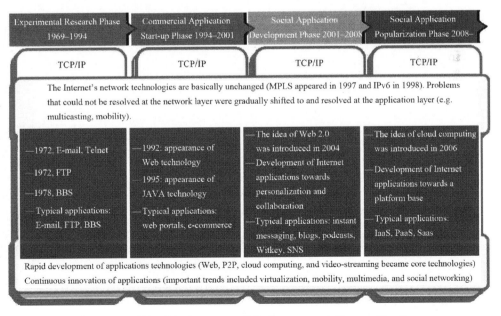

Experimental Research Phase 1969–1994	Commercial Application Start-up Phase 1994–2001	Social Application Development Phase 2001–2008	Social Application Popularization Phase 2008–
TCP/IP	TCP/IP	TCP/IP	TCP/IP

The Internet's network technologies are basically unchanged (MPLS appeared in 1997 and IPv6 in 1998). Problems that could not be resolved at the network layer were gradually shifted to and resolved at the application layer (e.g. multicasting, mobility).

		—The idea of Web 2.0 was introduced in 2004	—The idea of cloud computing was introduced in 2006
—1972, E-mail, Telnet —1972, FTP —1978, BBS Typical applications: E-mail, FTP, BBS	—1992: appearance of Web technology —1995: appearance of JAVA technology —Typical applications: web portals, e-commerce	—Development of Internet applications towards personalization and collaboration —Typical applications: instant messaging, blogs, podcasts, Witkey, SNS	—Development of Internet applications towards a platform base —Typical applications: IaaS, PaaS, Saas

Rapid development of applications technologies (Web, P2P, cloud computing, and video-streaming became core technologies)
Continuous innovation of applications (important trends included virtualization, mobility, multimedia, and social networking)

Figure 2-1: The development of the Internet and its applications

As the Internet became commercialized in the 1990s, it evolved from a network that was primarily used for scientific research to an information infrastructure on a global scale. With the rapid development of network technologies in recent years, there have been several new access technologies such as WiMAX, Wi-Fi, wireless local area networks, Bluetooth, and a large number of new networks with dissimilar structures such as mobile ad hoc networks, wireless sensor networks, and mesh networks. There have also been many new computing technologies such as P2P, grid computing, pervasive computing, and a variety of applications.

While these new technologies and applications promote progress in the field of communications, they also pose enormous challenges to the traditional architecture of the Internet. The complex environment of differently structured networks increases the complexity of maintenance. At the same time, the flexibility, robustness, and security of the network are affected. Ubiquitous application needs require the Internet to support mobility, while a diversity of applications requires it to support a variety of businesses in real time. With the increase in network complexity, the traditional Internet, which operates with a simple core, is unable to cater to the pressing needs of a controllable and manageable network. The Internet's infrastructure, with IP protocol and its corresponding addressing and routing mechanism as its core, has been operating for 40 years, and is increasingly overwhelmed. This poses serious challenges to its sustainable development.

All existing network architectures have their shortcomings when faced with current application needs. Reflecting on these issues is of great significance to future research into network architecture, and should be the starting point of further study.

1) Safety

It is generally accepted that traditional telecommunication networks are secure because their network hierarchy and boundaries are strictly defined. Users are only able to access the UNI, while the networks and applications of the NNI, SNI, and above are not directly accessible. However, with the Internet, end-to-end transparency and relatively flatter networks mean that users can access any network element, and are able to attack devices in the network. Therefore, network security is poor. At present, security risks exist in every aspect of the Internet, ranging from design and implementation to operation management. Frequent security breaches are a blatant manifestation of this issue. The growth of such incidents far exceeds the rate at which the scale of the Internet is growing, and security issues have become the main blockade to further development.

(1) Security was not given due consideration during the design stage of the Internet, and as such there is no security architecture for the system. There is no comprehensive

security mechanism built in under the TCP/IP protocols. As a result of resolving security problems as they arise, existing security technologies are a hodgepodge. Therefore, problems such as security vulnerabilities, overlapping functions, and implementation complexities are inevitable.

(2) There are many security vulnerabilities in the network operating system and the implementation of network protocols. They provide hackers with opportunities to attack.

(3) In the operation and management of networks today, there exist various security loopholes and incongruences between security mechanisms and management policies due to vulnerabilities in the implementation of computer software and hardware systems. Therefore, the security flaw in the Internet is a problem of architecture, which can only be completely resolved by making changes to the architecture itself.

In future research into network architecture, an important basis for testing the rationality of a network's architecture is whether it is able to provide security.

2) Mobility

For the Internet, ubiquity requires that the network can provide users with omnipresent, ubiquitous, and comprehensive services. The traditional Internet designed for fixed-location hosts and single data services is far from ubiquitous.

(1) The issue of mobility with the Internet has not been entirely resolved. Since the original TCP/IP protocol was designed for hosts with fixed locations, the IP address has two functions: firstly, to indicate the location of the host for routing at the network level; secondly, to identify the host for establishing a connection at the transport level. This kind of functional coupling means that dynamic binding of the host and the IP address cannot be supported, and thus the issue of the host's mobility cannot be fully resolved.

(2) Many smart mobile devices such as laptops and PDAs, and mobile communication devices such as mobile phones need access to Internet services. In future, many smart devices may also be connected to the Internet. However, at present, the Internet is still unable to support such a variety of access technologies.

(3) At present, the majority of Internet services revolve around the centralized client-server model, and fall short of being widely integrated and ubiquitous. When the traditional telecommunications network architecture was taking shape, the demand for mobile communication was not high, so the mobility of the terminal was not a consideration. Mobility was added to the telecommunications network later. Therefore, when using a top-down approach to study network architecture in the future, due consideration needs to be given to mobility and ubiquity.

3) Network performance

Within the current architecture of Internet protocols, both network performance and quality of service pose challenges. Essentially, the Internet provides "best effort", connectionless service. It simply tries to send packets to their destination as best as it can without providing any guarantees of service quality on either the network throughput, latency, or jitter. In a data service environment that primarily comprises FTP, E-mail, and Web services, the Internet is mostly able to meet users' needs. However, a large number of new applications, such as real-time streaming media transmissions like voice calls and IPTV, require high-quality network service. The Internet is unable to provide the performance and service quality required. As traditional telecommunication networks rely on signaling and are connection-oriented, a channel (or virtual circuit) would have been established between both ends and resources reserved prior to communication. Therefore, through the communication requirement of admission control, the quality of service can be guaranteed in the subsequent communications. It is necessary to consider network performance and quality of service from the network architecture perspective in order to fully resolve the underlying problems.

4) Controllability

A lack of controllability in the network is also a problem of Internet architecture. One of the core concepts of the Internet is to carry out memoryless transmission of data. State information is minimized or done away with so as to ensure the simplicity and efficiency of network equipment. This means that there is a lack of the information necessary for management in the network elements. As such, the administrator is unable to manage and control the network efficiently. At present, the Internet still relies on the original management techniques – a management system based on the level of the network elements. The system is complicated, difficult to operate, inefficient, and unable to adapt to today's huge and complex network systems. In telecommunication networks, major importance is placed on network controllability. As such, their network architecture already includes the collection and management of both network and status information. The industry has also formulated and developed a network management system. Therefore, the consensus is that in terms of network management, there is no deficiency in the architecture of the telecommunication network. Nonetheless, issues such as the management of entire networks (rather than individual devices), automated network management and configuration tools, robust and distributed network monitoring, error alerting and rapid discovery, and global user behavior-tracking and control are yet to be solved.

5) Credibility

The Internet lacks credibility mechanisms, and this can be attributed to its architecture. The openness and anonymity of the Internet mean that credibility is not guaranteed. This crisis of trust is manifested in incidents such as illegal intrusion, address spoofing, identity fraud, and cyber scams. For the vast majority of incidents, it is impossible to trace the perpetrators.

(1) During the initial design phase of the Internet, it was thought that its only users would be researchers, and thus trustworthy. As such, no thought was given to incorporating any credibility mechanism. Today, there are many types of Internet users, and there is no lack of malicious individuals who are bent on causing damage. This is the direct cause of the prevalence of security issues. Therefore, it is urgent to introduce an identity verification mechanism for Internet users to ensure a credible network environment.

(2) Existing routing devices on the Internet send packets based on the destination address. This means that the intermediate nodes in the network neither verify the source nor audit the transmitted data packets. As a result, address spoofing, spam, extensive intrusions, and attacks cannot be tracked.

(3) At present, a large number of netizens use private addresses to access the Internet via network address translation (NAT) making it extremely difficult to track security incidents.

(4) An end-to-end identity verification mechanism which can be used widely is unavailable at present. In its current form, the Internet lacks credibility. This places the numerous key business systems that are built on it in a precarious position, which impedes the informatization process.

Due to the strict regional division in telecommunication networks, there are UNI, NNI, and SNI access points. For a particular operator, NNI and above are considered credible. Similarly, the networks between different operators are also considered credible. As such, the general consensus is that the network architecture of traditional telecommunication networks is flawless in terms of credibility. The credibility mechanism of the network is one of the key issues that needs to be considered in network architecture, and is also an important basis for testing the rationality and feasibility of the architecture overall.

2.1.2 Rising global interest in researching the next-generation Internet

To solve the current problems that the Internet faces, and also to compete for the vantage point in future information technology, countries and regions such as the United States, Europe, China, Japan, and South Korea have embarked on research into the future of the Internet. Most of the goals of these research programs focus on a new type of network that can replace the existing Internet 10 to 15 years from now. Some of the most notable include:

- Future Internet-related projects such as 4WARD and FIRE in the EU FP7 (Seventh Framework Programme)
- The FIND, GENI, and PlanetLab projects supported by the US NSF;
- China's FPBN, hierarchical networks, and universal networks;
- Japan's AKARI, and JGN2+.

Currently, the United States and the European Union are the centers of global research into the next-generation Internet.

1) United States

With its first-mover advantage in the Internet field, the United States hopes to consolidate and strengthen its hegemony by maintaining the existing Internet governance structure and adhering to the existing technological system. On the other hand, the US also fully recognizes the problems afflicting the Internet today. To this end, it is stepping up efforts to roll out future networks ahead of schedule, and is strengthening the innovation and experimentation of future network technologies in the hope of extending the advantages it currently enjoys in the long-term evolution of the Internet.

The National Science Foundation (NSF) is responsible for research and experimentation in future networks. Since 2002, NSF has launched GENI (Global Environment for Network Innovations) and FIND (Future Internet Design). GENI focuses on the phased establishment of experimentation and verification platforms that support network architecture and key technological research. At present, an experimental network with more than 2,000 nodes worldwide has been established. Meanwhile, the FIND initiative has funded nearly 50 research projects related to future networks. FIND focuses on research and innovation in Internet network architecture and key technologies. It is also dedicated to basic research that may solve a series of fundamental problems in the system architecture of future networks, with more attention being paid to details and protocols.

In 2010, the NSF funded five key research projects on future networks: ChoiceNet, NEBULA, MobilityFirst, XIA, and NDN. It is gradually shifting the priorities of its research projects to multi-service support, cloud computing, mobility, and network infrastructure (name-based routing). These projects have an important effect on the direction in which future-network technologies are developing around the world. In terms of the standard of research, these projects are also world-leading. They have become the focus of follow-up studies in other countries, and also guide research into future-network technologies.

(1) ChoiceNet

ChoiceNet emphasizes the openness of the network and attempts to avail users of the functions of every layer of the protocol stack through the design of a new architecture. It has established an ecosystem that includes users, service providers, and developers, which enables users to choose the type of technology and service they require, thereby optimizing the economy of the network.

(2) NEBULA

NEBULA is geared towards the interconnection needs of data centers for cloud computing. It aims to create a completely new network architecture that brings future networks into the cloud model, enabling the management of resources and public pricing for the Internet.

(3) MobilityFirst

This project aims to facilitate network mobility by studying the establishment of generalized delay-tolerant networks (GDTN) to improve stability of communication. The focus is on striking a balance between mobility and scalability, and making full use of network resources to achieve effective communication between mobile endpoints.

(4) XIA

XIA is designed to address the diversity of network usage and the issue of trustworthy propagation. It also aims to enable network reconfiguration to ensure the flexibility, diversity and security of propagation paths.

(5) NDN

NDN is designed to promote efficient distribution of content. A new network architecture will be implemented through the redesigning of names, addresses, and routes. This will allow the Internet to deliver content directly, regardless of the physical location where it is stored.

In the United States, research into the Internet of the future is mainly driven by the NSF and leading universities. The FIND program, which began in 2005, originally funded nearly 50 research projects related to the Internet of the future. In 2010, the NSF shifted its research priorities to multi-service support, cloud computing, mobility, and network infrastructure (name-based routing). At the same time, the United States is also putting a lot of effort into the construction of testbeds. PlanetLab has set up more than 520 sites worldwide with 1,138 nodes. The strategic intention is to become the backbone infrastructure for the Internet of the future.

2) The European Union

The EU hopes to seize the opportunities arising from this new round of changes to network technologies. Instead of following the United States' lead, as it has been doing thus far, it hopes to overtake it in the Internet field. The EU's Seventh Framework Programme (FP7) was launched in January 2007. The program's Future Internet Research and Experimentation (FIRE) project sought to enhance research in Internet architecture and key technologies, and to construct a testbed for future networks. The goal is to establish an experimentation platform for Europe, to support research into new methodologies for tackling issues related to network scalability, complexity, mobility, security, and transparency.

Through the FP7, the EU has funded many research projects on the future of the Internet. These can be divided into six major areas: business, media, the Internet of Things, security, network architecture, and testbeds. The most important component is research into future Internet architecture and the setting up of testbeds. Investment into the testbed project in FIRE alone has reached EUR$40 million.

In terms of research into network architecture and key technologies, nearly a hundred projects are started each year. The priorities are research into transport-layer network technologies such as content-routing and P2P; technological innovation at the infrastructure level for applications such as CDN and IDC; and business innovation for typical applications such as cloud computing, mobile Internet, and the Internet of Things. The outcomes of these projects will be verified experimentally using the future network testbeds supported by FIRE.

In terms of setting up future network testbeds, special examples for specific technical solutions will be built, such as WISEBED (Wireless Sensor Network Testbeds), which provide multi-layer infrastructure for testing large-scale wireless-sensors. Meanwhile, existing testbed resources within the EU will be integrated to build general-purpose testbeds for the region. One example is the PII project, which looks into integrating the existing testbeds with those being built to achieve the sustainable long-term development of both dedicated and general-purpose beds. At present, FIRE testbeds in the EU have nearly 1,000 nodes, and are connected with the United States' GENI testbeds.

3) Japan

Japan attaches great importance to research into future networks, and has initiated a number of research projects to this end. The most prominent is the AKARI project, which was initiated by the National Institute of Information and Communication Technology (NCIT). The AKARI project aims to study the architecture and core technologies of next-generation networks to make up for the inadequacies of existing IP-based networks in security, mobility, and resource management. The project is divided into four levels: applications, coverage, IP, and base. It defines three core aspects of future networks:

- They are streamlined and intelligent;
- They are physically connected;
- They are sustainable and evolvable.

Since its launch in 2006, the project has made progress in areas such as parallel optical packet transmission theory, all-optical path/packet switching, packet multiple access, identity/location separation, and network virtualization. It is predicted that by 2016 the preliminary findings of the project will form the basic architecture of future networks.

4) South Korea

South Korea began research into new network architecture for the Internet in 2005. In 2010, it established the Korea Communications Commission and rolled out the Future Network Strategy. Its aim is for future networks to integrate communications, broadcasting, computing, and sensing networks. This would create an environment that consistently provides the best service anywhere, according to the user's status and characteristics. Doing so would break through the limitations of the existing network architecture and establish a technical-and-service model that is compatible with integrated services and multiple terminals. Its guidelines are to ensure that South Korea has technological capabilities in all fields such as service, terminals, and network architecture. Through tackling existing social problems and creating new markets, future networks will be both a "green" growth-driver for sustainable development, and a new driver for the development of South Korea's economy in the 21st century.

Figure 2-2 shows the scope of research into the future of the Internet in the US and the EU.

Looking at research methods in countries and regions such as the US and the EU, the basic idea is to proceed simultaneously on the two fronts of theoretical research and building experimental platforms. Theoretical exploration is carried out

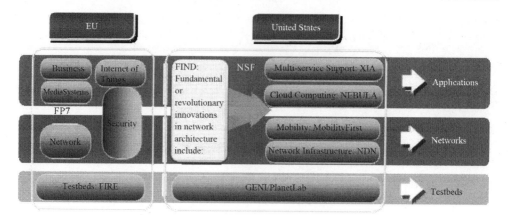

Figure 2-2: Scope of research into the future of the Internet in the US and the EU

in architectural innovation, exchange and transmission systems, and key algorithms. Meanwhile, experimental network platforms are built for the large-scale verification of the findings from theoretical research with the participation of actual users. There is much that China can learn from such an approach.

2.2 Developmental goals of the next-generation Internet

2.2.1 Core Challenges and Network Objectives for the Internet

1) Requirements of the next-generation Internet

At present, the Internet is a global information infrastructure. It has multiple roles, such as being a commercialized platform, a carrier of social information, a tool for mass communication, and new form of public media. Its industrial chain is also becoming more and more complicated, with each aspect making higher and higher demands.

- **Users**: Users need networks that are more secure. Terminals and client networks need to be kept safe, and the security of users' private information must be ensured. At the same time, increasing demands from users also pose higher requirements for the carrying capacity of network business, such as support for multicasting and mobility.
- **Network operators**: For network operators, the Internet is more than just a channel to deliver information. It should be a network resource that is flexible and can be easily managed and operated. At the same time, it should be a platform that allows the rapid expansion of new businesses.
- **Content providers**: For content providers, network quality (i.e. problems with users' perception of applications) is not the only issue of concern. It is

also necessary to improve the efficiency of resource utilization and to reduce operating costs.

- **Governments:** The Internet is a brand-new platform for social information and public opinion. Governments need to pay attention to the impact of the Internet on social life, and deal with new harmful issues such as cyber-crime and cyber-attacks. At the same time, the Internet boom also means that it is now a major energy-consumer in the world, and will become a focus of energy-saving and emission-reduction efforts in future.

2) Challenges faced by the next-generation Internet

The problems afflicting the Internet at present can be grouped into six major challenges:

- Depletion of IP addresses;
- Explosion of global routing tables;
- Lack of flexibility in business models;
- Serious cyber security issues;
- Inability to guarantee the quality of business applications;
- Growing energy consumption.

3) Key target needs of the next-generation Internet

The goal of the next-generation Internet is to solve the challenges that the existing Internet faces, and to provide society with a network platform that is: 1) able to support multiple business applications; b) safe and reliable; c) provides guaranteed quality of service; d) is rich in network resources and scalable; and d) meets the requirements of energy conservation. Based on this, the key target needs of the next-generation Internet can be summarized as:

- Sufficient IP addresses;
- Highly scalable routers at network layers;
- Supports a variety of business models and network structures;
- Is safe and reliable, manageable and controllable;
- Is able to provide guaranteed quality of service;
- Meets energy conservation needs.

Only network architectures that fully satisfy these needs are acceptable.

(1) Safety

Future network protocol architectures need to meet four basic security requirements:

- Confidentiality of information, to ensure that information cannot be accessed by unauthorized users;
- Integrity of data, to ensure that information is not altered during transmission;
- Authentication, so that a user is able to verify the identity of a correspondent;
- Non-repudiation, so that the user cannot deny sending a message after it has been sent.

Integrity of data means preventing information from being tampered with, while confidentiality prevents the leakage of information to unauthorized third parties. Availability refers to the prevention of unauthorized possession and preservation of resources or information, while authentication refers to the process of verifying the identity of the user who has logged in to the system. Authorization allows authorized users to access sensitive information and protected services or resources, and perform controlled operations. To achieve these goals, the next-generation Internet needs to establish a security framework. A security platform for detecting intrusions must be created, and a firewall must be set up to prevent viruses. This system will be made up of network components such as routers, wireless network controllers, gateways, and servers, and will ensure the integrity and reliability of devices for recovery from technical faults. In addition, security measures need to be added to the communication protocols to protect the data transmission channel, and to encrypt the data to protect the content.

Target characteristics of the next-generation Internet are shown in Figure 2-3.

Future networks must provide prevention prior to an attack, warning notifications during an attack, and audit trails following an attack, to achieve three-in-one integrated security. From a horizontal perspective, this will involve the division of the network security domain, a user-access security mechanism, and a security mechanism between the interior of the security domain and the security domain itself. In terms of the

Figure 2-3: Target characteristics of the next-generation Internet

composition of network functions, it will involve an isolation mechanism between the user plane and the network control and management plane.

(2) Mobility

The next-generation Internet needs to provide users with integrated round-the-clock services that are comprehensive and ubiquitous. It also needs to support the seamless and rapid movement of communication terminals, a wide range of terminal access technologies, and large-scale distributed and ubiquitous services so that the network is on the user's side.

(3) Network Performance

In future, carrier networks need to support high-bandwidth applications, and provide multimedia data transmission that is faster, smoother, and richer. They must be able to provide the quality of service that users need at a reasonable price, and should offer different network environments for a variety of services, which are personalized in terms of network performance. In future, carrier networks will need to support a diversity of services that include both "best effort" services and guaranteed services as and when necessary. Carrier networks will also need to support the classification of users' business types, priority level processing, service quality assurance mechanisms, and network-resource allocation and usage-management mechanisms, so as to provide services that fulfil the requirements of the service level agreement (SLA). This will help operators to build a more rational business model and allow users to enjoy the many conveniences of the network.

(4) Controllability

In future, network management needs to be simple and efficient to suit complex and large-scale network systems. Management and control of the entire network and not just of a single device is provided. Manual management, automated network management, and configuration tools are provided. Large-scale and distributed network monitoring, early warning and rapid detection of errors, and global user behavior-tracking and control are supported. Network operations, administration, and management (OAM) mechanisms that are convenient, accurate, and low-cost are provided. This means that service quality checks can be conducted without interrupting user services, thus providing the ability to locate network faults rapidly and accurately.

(5) Credibility

Firstly, future networks need to provide user-identity verification mechanisms to ensure that only credible users have access to the network or service. Secondly, future networks need to verify and audit the origin of transmitted data packets to prevent deceitful behavior such as address spoofing. Thirdly, future networks need to have a cross-domain global authentication system in place. Fourthly, a user-credibility model must be established to provide the basis for trusted applications. Finally, networks need to set a measure of credibility, and provide users with credible services. They must also require the various mechanisms of network security, control, and management to jointly achieve the credibility indicators.

(6) Network Transition
The architecture of future networks should be compatible with that of existing IP networks at certain levels to ensure coexistence, interoperability, and smooth transition.

(7) Supporting a variety of services
Future networks need to offer multiple services and provide network environments that are open, comprehensive, and flexible in order to support new businesses. Future networks should also be able to support the expansion of existing services and be accessible to new services by third parties. At the same time, they should support the rapid development and promotion of new services.

(8) Compatibility with underlying networks with heterogeneous structures
Future networks need to be backward-compatible with a variety of heterogeneous networks in order to become universal carrier networks. The compatibility of an IP network with the underlying heterogeneous network is the foundation of its successful application.

4) Evolutionary goals of future networks

The existing Internet is based on the TCP/IP protocol invented in the 1970s and '80s. The popularization and widespread use of the Internet exacerbates problems such as insufficient network addresses, a lack of a security and credibility, inconsistent service quality, a defective content-dissemination mechanism, and difficulties in network supervision. To a large extent, these problems hinder the development of the Internet. Since it is a strategic global information infrastructure, the next step in the evolution of the Internet is particularly crucial. The Internet is now at a critical period of technological change and evolution into the next generation.

The short-term evolutionary pathway for the Internet is relatively well-defined, and the understanding of this matter both in China and overseas is relatively uniform, i.e. IPv6 is the starting point of the process and is the foundation for the next-generation Internet. However, IPv6 is unable to solve all of the problems mentioned above, and is not a panacea for the next-generation Internet. In the medium to long term, the evolution of the Internet requires the enhancement of technological innovations and applications based on IPv6, to continuously tackle developmental issues and upgrade network capabilities.

Faced with the medium- to long-term evolutionary needs of the next-generation Internet, research is being done into the creation of a network architecture and infrastructure that has ample addresses and efficient distribution, and is safe, credible, ubiquitous, reliable, manageable, and controllable. In Europe, the United States, Japan, and South Korea, future networks are also known as the next-generation Internet or next-generation Internet. Even though the terms used differ, there is no significant difference in what they represent. They are efforts to explore the Internet's development and evolutionary pathway in the medium to long term based on needs such as scalability, security, mobility, assurance of service quality, and content distribution. "Future networks" is the name for the target network system of the medium- to long-term evolution of the next-generation Internet.

5) Evolutionary routes of future networks

In terms of the medium- to long-term development routes for the Internet, there currently a distinction made between the gradual route and the revolutionary route around the world. This can be applied to future networks as well.

The gradual route adopts an evolutionary mode of compatibility with existing networks, i.e. technological innovation based on the existing TCP/IP protocol, continuous improvement of the Internet's technological system, gradual solving of the existing problems with the Internet, and paying attention to the practicality of technological improvements. Its main feature is an evolution towards future networks through changing the routing control strategy based on the IPv6 data transfer format and the forwarding mode of connectionless packet-switching nodes.

The revolutionary route directly addresses the new needs without considering compatibility with the existing networks. Network infrastructure is redesigned to satisfy various fundamental needs. The revolutionary route focuses on advanced research in technology, and attempts to fundamentally change the TCP/IP system in search of a breakthrough. Japan's new-generation Internet aims to propose a new network system, and has positioned itself as the network for the next 15 years. Research projects in the United States such as FIND and FIA are named after the next-generation Internet. The subsequent GENI project clearly takes the revolutionary route. However, these research projects face tremendous difficulties, because it remains to be seen whether the findings from small, experimental networks can be replicated on large networks. Thus far, progress in these projects is way below expectations.

In the long-term evolution of the Internet, both the gradual and revolutionary routes are complementary and spur each other on. The ITU-T SG13 (next-generation Internet study group) leads the ITU's research on future networks technical standards. The group has completed and released the ITU-T Y.3001 Future Networks: Indicators and Design Goals. In it, "future networks" is defined rather broadly without explicitly including or excluding the TCP/IP technology system from future network research. The publication also mentions both brand-new network systems and improved network systems as possibilities, as well as the convergent development of new network architecture with existing networks. Therefore, any new network that aims to tackle the major technical challenges currently faced by the Internet, or to satisfy usage needs in the next 15 to 20 years can be called a future network regardless of whether it chooses the gradual or revolutionary approach. In fact, not all of the five FIA future network projects supported by the NSF in the United States have taken the revolutionary route. Some have selected the gradual route.

2.2.2 Component Model for Next-generation Internet Technology

From the perspective of technology, the development of the Internet's overall structure is a process that spreads from the core to the periphery, and from infrastructure to external demand.

Initially, the core technologies of the Internet were TCP/IP, as well as the routing technology closely associated to it, and mapping technology for resources and addresses. These are basic technologies that solve the naming, addressing, and routing problems of the network. With these basic problems out of the way, the IP network then constitutes a communication network that enables end-to-end data transfer.

As networks develop and the diversity of applications increases, both applications and network topologies are requesting more from IP networks. Besides satisfying basic communication needs, networks also need to expand their service support capabilities.

When a network has developed sufficiently to become the main information infrastructure, it will then face demands from all sectors of society to be safe and reliable, operable and manageable, and meet the requirements of energy saving. Even though these demands actually originate from outside the domain of network technology development, they still have a significant impact on it.

Summarizing the above analysis, the target characteristics of the next-generation Internet can be combined to obtain a technology component model for the Internet. This model should be based on the core architecture of the Internet (naming, addressing, routing, and the corresponding resource management) before gradually expanding to service support capabilities (such as terminal mobility, multicast or broadcast capabilities, and multiple user access). At the same time, it is necessary to

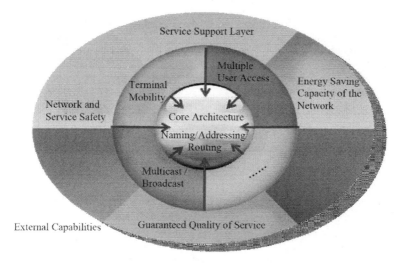

Figure 2-4: Internet Technology Component Model

support external demands from other sectors, such as safety, service quality, and energy saving, as shown in Figure 2-4.

Under this component model, the Internet's core architecture is the most critical and fundamental portion. The technical needs of all the other layers ultimately need to be reflected in the basic capabilities that can be provided by the core architecture, or for the core architecture to be altered or improved accordingly.

The current development of the Internet has two routes. Firstly, there is the improvisation route, which is basically an evolution of the existing Internet. By making use of improvisation techniques, network capabilities are improved without changing the core architecture of the Internet. Secondly, there is the revolutionary route, which looks at rebuilding the Internet's network infrastructure according to its demand model so that the new network can fundamentally support various external needs. In fact, the fundamental difference between the two approaches is whether the core architecture of the Internet is changed. In our view, this is also the most crucial distinction between the improvisation route and the revolutionary route.

2.2.3 The convergent development of the two approaches

Even though the two approaches have numerous fundamental differences, they are not completely exclusive or incompatible. No one has yet been able to clearly describe the future evolution of the Internet, but both approaches seem to be converging, as shown in Figure 2-5.

On the one hand, the target characteristics of the next-generation Internet according to the revolutionary route stem from the users and service needs of Internet at present. At the same time, the current Internet also provides a point of reference for research into networks and applications under the revolutionary route. On the other hand, research into the revolutionary route also provides many ideas for the improvement of the existing networks. For example, proposals to resolve the problem of routing scalability such as LISP (Locator/ID Separation Protocol) and HIP (Host Identity Protocol) have drawn on the findings of the layering of addresses and routing within the revolutionary route.

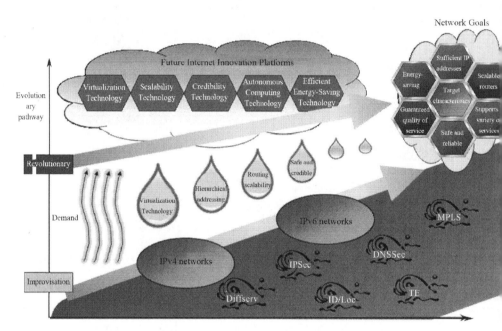

Figure 2-5: The convergent development of the two routes

Chapter 3

IPv6: Technical Features and Transition Mechanisms

Chapter Highlights:
* *Technical features*
* *Address format*
* *Header format*
* *Basic protocol*
* *Routing mechanism*
* *Network transition technologies*

Overview

With its abundant address resources, IPv6 solves the problem of insufficient IPv4 network addresses, and provides conditions for the improvement of network mobility and security. After more than 20 years of development and practice, IPv6 has become the sole solution for the next-generation Internet, and is relatively mature. Although it does not completely solve core issues such as network security and expansion, IPv6 network technology can gradually integrate new network technology concepts and technical elements for upgrading and evolution over the long term. As such, the industry considers IPv6 to be the starting point for the evolution of the next-generation Internet.

3.1 Technical features

As its name suggests, Internet Protocol version 6 (IPv6) is a new version of the Internet protocol. It was designed by the Internet Engineering Task Force (IETF) to replace the previous version (IPv4). The basic protocol is defined in IETF RFC 2460. The main differences between IPv6 and IPv4 are described below.

(1) Extended addressing capability
IPv6 increases the IP address size from 32 bits to 128 bits to support addressing hierarchy with more levels, a greater number of addressable nodes, and a simpler auto-configuration of addresses. IPv6 improves the scalability of multicast routing by adding a "scope" field to the multicast address. In IPv6, a new address type known as an "anycast address" is also defined, which is used for sending data packets to any node among a set.

(2) Simplified header format
IPv6 either omits some IPv4 header fields or makes them optional, thereby reducing the public processing overhead of data packets, and lowering the bandwidth overhead of IPv6 headers.

(3) Enhanced support for extensions and options
In IPv6, some changes have been made to the encoding of IP header options, aiming to make forwarding more efficient and to loosen restrictions on the length of the options, so as to provide greater flexibility for adding new options in future.

(4) Flow label capability
In order to satisfy any special handling requirements of the sender, IPv6 has added a new field to identify data packets belonging to a special transmission data stream, such as non-default quality of service or real-time services.

(5) Certification and confidentiality
IPv6 specifies extensions of functions that provide for authentication, data integrity, and data encryption (optional).
    Compared with IPv4, IPv6 has the following advantages.

- A larger address.
  In IPv4, the length of the IP address is 32 bits and the maximum number of addresses possible is 2^{32}. In comparison, the length of the IP address in IPv6 is 128 bits which means that the maximum number of addresses is 2^{128}. Compared with the 32-bit address size, the number of addresses has increased by $2^{128} - 2^{32}$. At present, IPv4 uses a 32-bit address size to generate some 4.3 billion addresses. By using a 128-bit address size, IPv6 can generate unlimited addresses,

thus providing sufficient address resources. Having rich address resources will completely eliminate many of the limitations that IPv4 Internet applications face, such as IP addresses. Every phone and every electrical device can have its own IP address. This would allow digital households to become a reality. At present, the technical advantage of IPv6 has to a certain extent solved the existing problems of the IPv4 Internet. This is one of the major drivers for the evolution from IPv4 to IPv6.

- A smaller routing table.
  From the onset, IPv6 addresses are grouped according to the principles of clustering. This enables the router to use a record in the routing table to represent a subnet, greatly reducing the length of the routing table in the router and increasing the speed at which the router forwards the data packet.

- Added support for automatic configuration.
  This is an improvement and extension to the DHCP protocol, making the management of networks (especially LAN) faster and more convenient.

- More secure.
  In networks using IPv6, users can encrypt data at the network layer and verify IP datagrams. The encryption and authentication options in IPv6 ensure the privacy and integrity of data packets, which greatly enhances network security.

- Room for expansion.
  IPv6 caters for the extension of protocols if new technologies or applications are required.

- Better header format.
  IPv6 uses a new header format with options that are separate from the basic header. The options can be inserted between the basic header and the upper layer data if needed. This simplifies and speeds up the routing selection process because most of the options do not need to be chosen by the router.

3.2 IPv6 address formats

The IETF RFC 4291 specifies the IPv6 address format. An IPv6 address is a 128-bit identifier assigned to an interface or group of interfaces. There are three types of IPv6 addresses, as shown below.

- **Unicast address:** The identifier of a single interface. The data packet sent to the unicast address is delivered to the interface indicated by this address.
- **Anycast address:** The identifier of a group of interfaces (usually belonging to different nodes). The data packet sent to the anycast address is delivered to one of a group of interfaces indicated by the address (the nearest one by distance as calculated using the routing protocol).
- **Multicast address:** The identifier of a group of interfaces (usually belonging to different nodes). The data packet sent to the multicast address is delivered to all the interfaces indicated by this address.

IPv6 does not use broadcast addresses. The function of broadcast addresses is replaced by multicast addresses.

3.2.1 Address Model

All types of IPv6 addresses are assigned to the interfaces and not the nodes. An IPv6 unicast address corresponds to a single interface. Since each interface belongs to a single node, the unicast address of any interface of the node can be used as an identifier of that node.

All interfaces need at least one link-local unicast address. A single interface may be assigned multiple IPv6 addresses of any type or range at the same time. An interface that is not the source or destination of any IPv6 data packet does not need a unicast address beyond the range of the link. At times, this can be useful for node-to-node interfaces. There is an exception to this address model. If several physical interfaces are treated as one during execution, when presented to the network layer, this may result in multiple physical interfaces being assigned a unicast or a group of unicast addresses. This is useful for load-sharing across multiple physical interfaces.

IPv6 has inherited the IPv4 model of subnet prefixes being associated with a link. Multiple subnet prefixes may belong to the same link.

3.2.2 The grammar of IPv6 addresses

IPv6 addresses come in three general forms.

The preferred form is x: x: x: x: x: x: x: x, where x refers to the eight hexadecimal 16-bit blocks in the address. For example:

FEDC: BA98: 7654: 3210: FEDC: BA98: 7654: 3210

1080: 0: 0: 0: 8: 800: 200C: 417A

The leading zeros in each block of the address do not need to be written, but there must be at least one number in each block (with the following exceptions).

Due to the varying methods of assigning different types of IPv6 addresses, the address usually contains a long string of consecutive 0s. For ease of writing an address of this form, there is a specific grammatical rule in RFC 4291 to abbreviate consecutive 0s. ": :" is used to replace one or multiple 16-bit blocks of 0s.

However, ": :" can only be used once in an address. ": :" can also be used to replace the consecutive 0s at the start or at the end of the address. For example:

1080: 0: 0: 0: 8: 800: 200C: 417A   Unicast address
FF01: 0: 0: 0: 0: 0: 0: 101   Multicast address
0: 0: 0: 0: 0: 0: 0: 1   Loopback address
0: 0: 0: 0: 0: 0: 0: 0   Unspecified address

They can be replaced by the following forms:

1080 :: 8: 800: 200C: 417A   Unicast address
FF01 :: 101   Multicast address
:: 1   Loopback address
::   Unspecified address

In environments with both IPv6 and IPv4 nodes, an address format of x:x:x:x:x:x:d.d.d.d can be used, where "x" is a hexadecimal number that is used in the most significant bits of the address (six 16-bit blocks), and "d" is a decimal number that is used in the less significant bits of the address (four 8-bit blocks). For example,

0: 0: 0: 0: 0: 0: 13.1.68.3
0: 0: 0: 0: 0: FFFF: 129.144.52.38

Alternatively, the abbreviated form would be:

:: 13.1.68.3
:: FFFF: 129.144.52.38

3.2.3 The grammar of address prefixes

The grammar of IPv6 address prefixes is similar to writing IPv4 address prefixes in CIDR notation. An IPv6 address prefix is represented in the form below.

IPv6 address/length of prefix

The IPv6 address referred to here can be any of the forms mentioned in Section 3.2.2. The length of the prefix is a decimal value that indicates the number of adjacent leftmost digits forming the prefix in the address.

The following example shows the proper ways of representing 12AB00000000CD3, a 60-bit prefix using hexadecimals:

12AB:0000:0000:CD30:0000:0000:0000:0000/60

12AB::CD30:0:0:0:0/60

12AB:0:0:CD30::/60

The following ways are incorrect:

12AB:0:0:CD3/60   It may lose the 0s in front but not those at the back.

12AB::CD30/60 "/"   The address on the left may be interpreted as 12AB: 0000:0000:0000:0000:000:0000:CD30 instead.

12AB::CD3/60 "/"   The address on the left may be interpreted as 12AB: 0000:0000:0000:0000:000:0000:0CD3 instead.

When writing the address of a node and the prefix of the address (such as the prefix of the node's subnet) at the same time, the two can be combined in the following form:

Node Address: 12AB:0:0:CD30:123:4567:89AB:CDEF.

Subnet ID: 12AB:0:0:CD30::/60.

Which can be abbreviated to: 12AB:0:0:CD30:123:4567:89AB:CDEF/60.

3.2.4 Identification of Address Types

The leading bits in the address indicate the type of IPv6 address. The leading bits of variable length are called format prefixes. Table 3-1 lists how the prefixes are allocated.

Anycast addresses are taken from the unicast address space of any range. Syntactically, it is difficult to separate anycast addresses from unicast addresses.

Due to other considerations, future standards may redefine one or more subspaces in the global unicast space, but unless and until such redefinition occurs, any address that does not have any of the prefixes mentioned above should be treated as a global unicast address in practice.

Table 3-1: Allocation of address prefixes

Address Type	Binary Prefixes	IPv6 Symbolic Notation
Unspecified	00...0 (128 bits)	::/128
Loopback	00...1 (128 bits)	::1/128
Multicast	11111111	FF00::/8
Link-local	1111111010	FE80::/10
Global unicast	Others	

3.2.5 Unicast Addresses

IPv6 unicast addresses are contiguous, maskable addresses that are measured in bits. They are similar to IPv4 addresses with CIDR.

In IPv6, there are the following types of unicast addresses: global unicast addresses, site-local unicast addresses (no longer in use), and link-local unicast addresses. There are also some purpose-specific subtypes of global unicast addresses such as IPv6 addresses with IPv4 address embedded. In future, newly-added address types or subtypes can also be defined.

The extent of an IPv6 node's knowledge of the internal structure of an IPv6 address depends on its function. In the simplest case, a node may treat an IPv6 address as a 128-bit string, as shown in Figure 3-1.

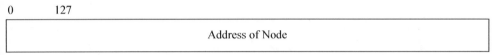

Figure 3-1: Internal structure of an IPv6 address (1)

In a slightly more complex scenario, a node may split the structure of an IPv6 address into two parts according to the prefix representing a subnet. The two parts are the network prefix and the interface ID, as shown in Figure 3-2. For different addresses, n may have different values.

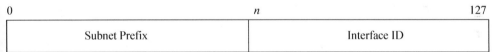

Figure 3-2: Internal structure of an IPv6 address (2)

For unicast addresses, more complex nodes can be classified in other ways. Routers can have a more specific understanding of the boundaries of single-level or multi-level structures. The extent of understanding depends on the location of the router in the routing structure.

1) Interface ID

In IPv6 unicast addresses, the interface ID is used to distinguish the interfaces in a link. They must be unique within the same subnet prefix. It is not recommended to assign the same interface ID to different nodes on the same link. An interface ID may also be unique in a broader range. In some cases, the interface ID can be obtained from the link layer address of the interface. Multiple interfaces on the same node can use the same interface ID as long as they belong to different subnets.

It is worth noting that the use of the same interface ID by different nodes neither affects the global uniqueness of the interface nor the global uniqueness of each IPv6 address that uses the interface ID.

For all unicast addresses, other than those that begin with the binary numbers 000, the interface ID has to be 64 bits long, and constructed according to the improvised EUI-64 format.

An interface ID based on the improvised EUI-64 format and originating from a global label (such as IEEE 802 48-bit MAC or IEEE EUI-64) can have global scope. When it is either unable to obtain the global label (e.g. tandem links or tunnel endpoints) or not interested in using the global label (such as for a temporary private label), an interface ID based on the improvised EUI-64 format can have local scope.

When an interface ID is formed according to the IEEE EUI-64 identifiers, an improvised EUI-64 format interface ID can be formed by inserting the "u" bit (universal/local bit, IEEE EUI-64 terminology). In the improvised EUI-64 format, a "1" in the location of the "u" bit indicates global scope, whereas a "0" indicates local scope. Figure 3-3 shows the first three bytes of the binary IEEE-EUI-64 identifier.

0		7 8		15 16	23
cccc	ccug	cccc	cccc	cccc	cccc

Figure 3-3: The first three bytes of the binary IEEE-EUI-64 identifier

Figure 3-3 shows the bit sequence written in accordance with Internet standards, where "u" stands for the universal/local bit, "g" stands for the individual/group bit and "c" stands for the company_id bits.

The IPv6 node does not need to verify the uniqueness of the interface ID generated using the improvised EUI-64 identifier (with the "u" bit set to global scope).

2) Unspecified addresses

The address 0:0:0:0:0:0:0:0 is called an unspecified address and is not assigned to any node. For a node that does not know its address in the initial state, this can be treated as the source address of the data packet.

An unspecified address can neither be the destination address of an IPv6 header nor the routing header of an IPv6 header. Routers are unable to forward IPv6 data packets whose source addresses are unspecified.

3) Loopback addresses

The unicast address 0:0:0:0:0:0:0:1 is known as the loopback address. A node can use it to send IPv6 data packets to itself. It cannot be assigned to any physical interface, and is considered to belong to the link-local scope. It can be treated as the link-local unicast address for a virtual interface (typically known as a "loopback interface"). The virtual interface leads to an imaginary link that is disconnected.

The loopback address cannot be used as the source address for IPv6 data packets that are sent from a single node to the outside. IPv6 data packets destined for the loopback address cannot be sent out of the node and cannot be forwarded by an IPv6 router. Data packets destined for the loopback address must be discarded by the interface.

4) Global Unicast Addresses

Figure 3-4 shows the general format of an IPv6 global unicast address.

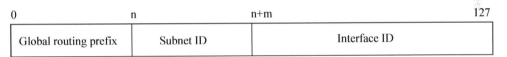

0	n	n+m	127
Global routing prefix	Subnet ID	Interface ID	

Figure 3-4: General format of a global unicast IPv6 address

In Figure 3-4, the global routing prefix is a value (typically hierarchical) assigned to a site (a group of subnets/links). The subnet ID is the identifier of the link within the site, whereas the meaning of the interface ID is consistent with that mentioned previously.

Other than global unicast addresses that begin with the binary numbers "000", all other global unicast addresses have a 64-bit interface ID field (i.e. n+m=64). Global unicast addresses that begin with the binary numbers "000" do not face such a restriction in terms of address size or interface ID field structure.

An example of a global unicast address that begins with the binary numbers "000" is an IPv6 address with an IPv4 address embedded.

5) IPv6 addresses with an embedded IPv4 address

Currently, the IETF defines two types of IPv6 address types for carrying IPv4 address information in the last 32 bits of the address, namely, an IPv4-compatible IPv6 address and an IPv4-mapped IPv6 address.

IPv4-compatible IPv6 addresses are used for the transition to IPv6. Figure 3-5 shows the format of an IPv4-compatible IPv6 address.

Figure 3-5: IPv4-compatible IPv6 address

The IPv4 address used in an IPv4-compatible IPv6 addresses must be a globally unique IPv4 unicast address. IPv4-compatible IPv6 addresses are outdated because current IPv6 transition mechanisms no longer use them. Therefore, new implementations are not required to support this type of address at present.

An IPv4-mapped IPv6 address is used to express the address of an IPv4 node as an IPv6 address. Figure 3-6 shows the format of IPv4-mapped IPv6 address.

Figure 3-6: IPv4-mapped IPv6 address

6) Link-local IPv6 unicast addresses

Link-local addresses are used for single link. Figure 3-7 shows the link-local address format.

0	10	64	127
1111111010	0	Interface ID	

Figure 3-7: Link-local address format

The link-local address is designed for addressing on a single link. It is used for automatic address configuration, neighbor discovery, or when there is no router on the link.

Routers cannot forward any data packet that has a link-local source address or a link-local destination address to other links.

7) Site-local IPv6 unicast addresses

Initially, the site-local address was designed for intra-site addressing that does not require a global prefix. It is now outdated. Figure 3-8 shows the format of the site-local address.

0	10	64	127
1111111010	Subnet ID	Interface ID	

Figure 3-8: Site-local address format

In new implementations, the special nature of this prefix as defined by the IETF RFC 3513 is not supported (i.e. the new implementation needs to treat this prefix as a global unicast).

This prefix can continue to be used in existing implementations and deployments.

3.2.6 Anycast addresses

An anycast address can be simultaneously assigned to multiple network interfaces belonging to different nodes. The significant feature is that a data packet with an anycast address as its destination address is forwarded to the nearest interface as measured according to the routing protocol. Anycast addresses are carved out from unicast addresses. Any of the defined unicast address formats can be utilized. Therefore, anycast addresses are grammatically indistinguishable from unicast addresses. When a unicast address is configured on multiple interfaces, the unicast address becomes an anycast address. At the same time, the node must be explicitly configured to know that the address in question is an anycast address.

Any arbitrarily assigned anycast addresses have the longest address prefix "P". This address prefix identifies the topological area where all the interfaces belonging to the same anycast address are located. Within the area indicated by "P", each member of an anycast address must be broadcast in the routing system as a single entity (usually referenced as a host route). Outside that area, the anycast address must be aggregated according to the address prefix "P" in the route broadcast.

It should be pointed out that in the worst case, the address prefix of an anycast address set may be an empty prefix. That is to say that members of the set may not have a local topological area. In this case, the anycast address must be broadcast throughout the Internet as a separate routing entry. This sets strict requirements on how many "global" anycast addresses the Internet should support. Therefore, it is foreseeable that support for "global" anycast addresses may not be feasible or only possible under very strict conditions.

One of the expected uses of an anycast address is to identify a set of routers that belongs to an Internet service provider. The anycast address can be used as an intermediate address in the IPv6 routing header to force the data packet to be forwarded through a specific aggregation or series of aggregations.

Some other possible uses of an anycast address include identifying the set of routers connected to a particular subnet, and identifying the set of routers that provides access to a particular routing domain.

The anycast address of a subnet router is predefined, and its address format is shown in Figure 3-9.

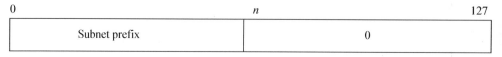

Figure 3-9: Anycast address format

The subnet prefix in an anycast address is a prefix that identifies a particular link. This anycast address is equivalent to a unicast address on the link whose interface ID is set to zero.

A data packet whose destination address is the anycast address of a subnet router will be forwarded to a router on that subnet. All routers are required to support subnet router anycast addresses that correspond to the subnet where their interfaces are located.

The purpose of the subnet router anycast address is for applications that need to communicate with a router in a set of routers in a remote subnet.

3.2.7 Multicast addresses

An IPv6 multicast address is used to identify a group of nodes. A node can belong to multiple multicast address groups. A multicast address has the format shown in Figure 3-10.

Figure 3-10: Multicast address format

11111111 at the beginning of the format indicates that this is a multicast address. There are four flags in the flgs field: | 0 | R | P | T |.

The flag of the most significant digit is reserved, and must be set to zero at initialization.

If the T flag is 0, this means that the address is a permanently assigned (well-known) multicast address, and is centrally assigned by the global Internet authorities. If the T flag is 1, it means that the address is a non-permanent (either temporary or dynamically-assigned) multicast address.

If the P flag is 0, the multicast address is not related to the network prefix. If the P flag is 1, it means that the multicast address is assigned based on the network prefix. In this case, the T flag must be set to 1 as well. When the P flag is set to 1, the meaning of the multicast address is as shown in Figure 3-11.

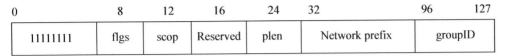

Figure 3-11: Multicast address format with P flag set to 1

The "plen" field indicates the actual number of effective bits in the network prefix field.

The network prefix field contains the network prefix of the unicast subnet to which the multicast address belongs. Meaningless bits in this field should be set to 0, and the maximum length of the network prefix in this field is 64 bits.

If the R flag is 1, it indicates that the multicast address has a multicast rendezvous point (RP) address embedded. In this case, both the P and T flags must be set to 1, and the multicast address prefix is FF70::/12. The last four bits of the 8-bit reserved field as shown in Figure 3-12 is the interface ID of the embedded RP. IETF RFC3956 specifies the method of generating an RP address. Figure 3-12 shows the RP address generated.

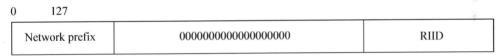

Figure 3-12: Format of an RP address generated using a multicast address

The value of the network prefix field is the first "plen" bits of the network prefix field in Fig. 3-11, while the RIID field is the last 4 bits of the reserved field in Figure 3-11.

The 4-bit-long scope determines the effective range of the multicast address. The details are as follows:

- 0 reserved;
- 1 interface local scope;
- 2 link-local scope;
- 3 unassigned;

- 4 management of local scope;
- 5 site-local scope;
- 6 unassigned;
- 7 unassigned;
- 8 organization local scope;
- 9 unassigned;
- A unassigned;
- B unassigned;
- C unassigned;
- D unassigned;
- E global scope;
- F reserved.

The group ID is used to identify the permanence or transience of the multicast address within its specified range.

The permanently allocated multicast address is independent of the valid range of the address. For example, if an NTP server group is assigned a permanent multicast address and its group ID is 101, then:

- FF01: 0: 0: 0: 0: 0: 0: 101 refers to all the NTP servers on the same node as the sender;
- FF02: 0: 0: 0: 0: 0: 0: 101 refers to all the NTP servers on the same link as the sender;
- FF05: 0: 0: 0: 0: 0: 0: 101 refers to all the NTP servers on the same site as the sender;
- FF0E: 0: 0: 0: 0: 0: 0: 101 refers to all the NTP servers on the Internet.

Multicast addresses that are not permanently assigned are only valid within the specified range. For example, the non-permanent multicast address FF15: 0: 0: 0: 0: 0: 0: 101 in a certain site is unrelated to a multicast address group that has the same address in other sites. It is also unrelated to non-permanent multicast address groups with the same group ID but different valid ranges, and is unrelated to permanent multicast address groups with the same group ID.

The multicast address can neither be used in the source address field of an IPv6 data packet nor in any routing header.

Routers are unable to forward any multicast data packet beyond the range indicated by the scope field in the destination multicast address.

Nodes are unable to generate data packets that contain the reserved value 0 in the scope field of the multicast address. A node should discard any such data packets received. Nodes are unable to generate data packets that contain the reserved value F in the scope field of the multicast address. If such a data packet is sent or received, the packet must be processed in the same way as a one with E (global scope) in the scope field of the multicast address.

The following is a list of pre-defined and well-known multicast addresses.

Reserved multicast addresses:

FF00: 0: 0: 0: 0: 0: 0: 0
FF01: 0: 0: 0: 0: 0: 0: 0
FF02: 0: 0: 0: 0: 0: 0: 0
FF03: 0: 0: 0: 0: 0: 0: 0
FF04: 0: 0: 0: 0: 0: 0: 0
FF05: 0: 0: 0: 0: 0: 0: 0
FF06: 0: 0: 0: 0: 0: 0: 0
FF07: 0: 0: 0: 0: 0: 0: 0
FF08: 0: 0: 0: 0: 0: 0: 0
FF09: 0: 0: 0: 0: 0: 0: 0
FF0A: 0: 0: 0: 0: 0: 0: 0
FF0B: 0: 0: 0: 0: 0: 0: 0
FF0C: 0: 0: 0: 0: 0: 0: 0
FF0D: 0: 0: 0: 0: 0: 0: 0
FF0E: 0: 0: 0: 0: 0: 0: 0
FF0F: 0: 0: 0: 0: 0: 0: 0

The above are reserved multicast addresses and must never be assigned to any multicast group.

- All node addresses:
  FF01: 0: 0: 0: 0: 0: 0: 1
  FF02: 0: 0: 0: 0: 0: 0: 1

The multicast addresses above identify all IPv6 node groups. The valid range is either the local interface scope or the link-local scope.

- All router addresses:
  FF01: 0: 0: 0: 0: 0: 0: 2

FF02: 0: 0: 0: 0: 0: 0: 2
FF05: 0: 0: 0: 0: 0: 0: 2

The multicast addresses above identify all IPv6 router address groups. The valid range is the local interface scope, link-local scope, or site-local scope.

- Solicited node address:
  FF02: 0: 0: 0: 0: 1: FFXX: XXXX

This multicast address is calculated based on the function of the unicast and anycast addresses of the node. It consists of two parts. One part is from the lower 24 bits of the address (unicast or anycast) concatenated with the prefix FF02: 0: 0: 0: 0: 1: FF00 :: / 104 to form a 128-bit multicast address. The address range is from FF02: 0: 0: 0: 0: 1: FF00: 0000 to FF02: 0: 0: 0: 0: 1: FFFF: FFFF, for example, the multicast address of the requesting node corresponding to the address 4037:: 01: 800: 200E: 8C6C is FF02 :: 1: FF0E: 8C6C. In this way, IPv6 addresses with different aggregation needs or that differ in the most significant bits will be mapped to the same solicited node address, reducing the number of multicast addresses linked with the node.

The node must calculate and join the corresponding solicited-node multicast address for each unicast and anycast address to which it is assigned.

3.3 IPv6 Header Format

Figure 3-13 shows the IPv6 header format.

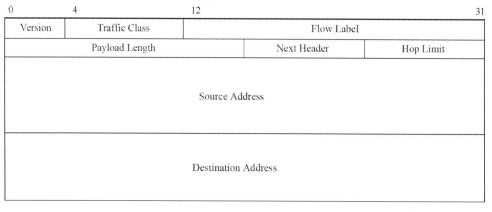

Figure 3-13: IPv6 header format

The meaning of each field in the header is as follows:

- **Version**: This field is 4 bits long and the version number for IPv6 is "6";
- **Traffic** Class: This field is 8 bits long. See Chapter 9;
- **Flow Label**: This field is 20 bits long. See Chapter 8;
- **Payload Length**: This field is a 16-bit unsigned integer indicating the IPv6 payload length i.e. the length of the rest of the data packet after the IPv6 header, measured in bytes. (Note: Any extension header is considered as part of the payload, and its length should be included);
- **Next Header**: This field is 8 bits long and indicates the type of header immediately following the IPv6 header. This field takes different values according to the type of extension header. For the correspondence between values and extension header types, please refer to the website for IANA protocol numbers and assigned services;
- **Hop Limit**: This field is an 8-bit unsigned integer. Each time the data packet passes through one forwarding node, the hop limit decreases by 1. When the hop limit is reduced to 0, the data packet is discarded;
- **Source Address**: This field is 128 bits long and represents the IPv6 address of the node that generated the data packet;
- **Destination Address**: This field is 128 bits long and represents the IPv6 address at which the data packet is expected to arrive. If a routing header appears, then this address may not be the IPv6 address of the final destination of the data packet.

1) The length of an IPv6 data packet

IPv6 requires that the MTU of any link on the Internet be no less than 1280 bytes. On any link that does not support 1280-byte data packets, a link-related fragmentation and reassembly function must be available at a layer below the IPv6 layer.

MTU configurable links must be configured with an MTU of at least 1280 bytes. For such links, the recommended configuration of MTU is 1500 bytes or more, to accommodate larger data encapsulation without fragmentation.

A node must be able to receive data packets as large as the MTU of any link to which it is directly connected.

IPv6 strongly recommends that "path MTU discovery" be implemented in nodes so as to discover and utilize paths with an MTU greater than 1280 bytes. However, a simple IPv6 implementation can restrict itself not to send data packets larger than 1280 bytes, thus leaving out "path MTU discovery".

To send data packets larger than the path MTU, the node must fragment the data packet using the IPv6 fragmentation header at the data source. The data packet then needs to be reassembled at the destination. However, the use of fragmentation (i.e. a minimum of 1280 bytes) is not recommended if the application can adjust its data packets to conform to the MTU of the path.

A node must be able to receive fragmented data packets that are 1500 bytes in size after reassembly. A node allows the receipt of fragmented data packets that are 1500 bytes in size after reassembly. An upper-layer protocol or application that relies on IPv6 fragmentation to send data packets longer than the MTU of the path should not send data packets longer than 1500 bytes unless it explicitly knows that the destination node is capable of reassembling such a large data packet.

When an IPv6 data packet is sent to an IPv4 destination, the initiating IPv6 node may receive a "data packet too large" ICMP message to report that the MTU for the next hop is less than 1280 bytes. In such a scenario, it is not necessary for the IPv6 node to reduce the length of subsequent packets to fewer than 1280 bytes. Instead, a fragmentation header must be added to the data packet so that the router that undertakes IPv6-IPv4 protocol translation is able to obtain the appropriate flag value to carry out IPv4 fragmentation. Note: this means that the payload length may be reduced to 1232 bytes (1280 bytes minus 40 bytes for the IPv6 header and another 8 bytes for the fragmentation header), or less if additional extension headers exist.

2) Flow label

The flow label field in the IPv6 header occupies 20 bits, and the source node uses it to mark a series of data packets that require special handling by an IPv6 router. This special handling includes non-default quality of service or real-time services. The content of this aspect was still in the experimental phase when IPv6 was formulated. As the direction in which the Internet develops becomes clearer, the content regarding the flow support requirements will be updated. For hosts and routers that do not support the functions of the flow label field, the field is set to zero when sending a data packet; no change is made to the field when forwarding a data packet, and the field is ignored when receiving the data packet.

A flow refers to a group of data packets sent from a specific source address to a specific destination address (unicast or multicast address) for which the source node requires that the flow be specially handled by the intermediate routers. The nature of the special handling can be conveyed to the router either by a control protocol (such as the resource reservation protocol), or through the information in the stream of data packets (such as hop-by-hop options). The details of such control protocols or options are beyond the scope of this standard.

There may be multiple active flows between a pair of sources and destinations. There can also be traffic that does not belong to any flow. A flow is uniquely identified by a combination of the source address and a non-zero flow label. Data packets that do not belong to any flow have a flow label of zero.

The label of a flow is specified by the source node. New flow labels must be chosen uniquely and pseudo-randomly from the range 1 to 0xFFFFF (hexadecimal). The purpose of random assignment is to make sure that any set of bits in the flow label generated are suitable for use as a key value in the hash table for routers. The hash key value is used to find the state that corresponds to the flow.

All data packets belonging to the same flow must have the same source address, destination address, and flow label when sent. If any of these data packets contain a hop-by-hop option header, then each packet in the flow must contain the same one (excluding the next header field in the hop-by-hop option header). If any of these data packets contain a routing header, then each packet in the flow must contain the same one, including the extension header(s) of the routing header (excluding the next header field in the routing header). The routers or destination nodes are allowed but not required to verify whether these conditions are met. If a data source that doesn't meet these conditions is detected, an "Parameter Error" ICMP message (Code 0) should be sent to the source node. The error message points to the most significant byte of the flow label, i.e. the second byte of the IPv6 packet.

The maximum lifespan of the flow processing state established along the flow path needs to be specified in the state establishment mechanism, such as in the resource reservation protocol, or in establishing the hop-by-hop options for the flow. The source node will not allow this flow label to be assigned to a new flow within the maximum lifespan of any flow processing state, because the state might have been established before the flow label is used.

When a node reboots, such as resuming operation after a crash, flow labels must be used with great caution, as a particular flow label might still be in use by a prior flow that is still within its maximum lifespan. This can be accomplished by recording the usage of the flow label on static storage so that the information is preserved after a crash recovery, or by avoiding using any flow label before the maximum lifetime of any previously established flow has expired. If the minimum reboot time of a node is known, the actual reboot time can be derived from the time it takes to wait for the allocation of a flow label.

It is not required that all or even most of the data packets (i.e. those with non-zero flow labels) belong to a certain flow. This rule reminds protocol designers and implementers not to assume the opposite. For example, the router only performs well when most of the data packets belong to a certain flow, or if the router's header compression mechanism only processes data packets that belong to a flow. Designing a router in these ways is not reasonable.

3) Traffic class

The 8-bit traffic class field in the IPv6 header is used by the source node and/or router to determine the class or priority of the IPv6 data packet. Currently, it has been discovered through testing IPv4 that the traffic type field can be used to provide IP data packets with differentiated services. The traffic class field in the IPv6 header has similar functions.

Listed below are some general requirements of the traffic class field:

- The IPv6 service interface must provide upper-layer protocols with a means of modifying the value of the traffic class field when generating data packets. The default value for this field is 0;
- Special application nodes that support the traffic class field are able to change the value of this field according to the requirements of the special application when generating, forwarding, or receiving the data packet. Nodes that do not have this capability should ignore the field, and should not be allowed to modify it;
- The value of the traffic class field in a data packet received by an upper-layer protocol may differ from the value of the traffic class field in a data packet sent by the source node.

4) IPv6 extension headers

In IPv6, optional Internet-layer information is encoded in separate headers and placed between the IPv6 header and the upper-layer header of a data packet. The number of such extension headers is small, and is clearly defined by the value of a "next header" field. As shown in Figure 3-14, each IPv6 data packet can carry zero, or one, or more extension headers, each of which is defined by the "next header" field of the previous header.

Other than the "hop-by-hop option" extension header, all other extension headers are not checked or processed by any node in the data packet's delivery path until it reaches the node specified in the "destination address" field in the IPv6 header (or in the case of multicast routing, any node in the group). An exception is defined in the IETF RFC 7045. If a forwarding node needs to examine extension headers for certain purpose (e.g. a firewall), then the node should be able to identify and process the extension headers defined in the IETF. IETF RFC 2460 requires destination nodes to discard data packets containing unknown extension headers. However, forwarding

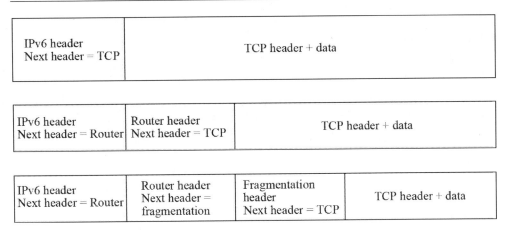

Figure 3-14: IPv6 extension headers

nodes along the path are forbidden to do so, as the unknown extension headers may be newly-defined, and the implementation of intermediate nodes may not have been upgraded to the most up-to-date version. Intermediate nodes can only discard data packets based on preconfigured policy but not because they contain unknown extension headers. Intermediate nodes need to be able to be configured to allow the forwarding of data packets containing unknown extension headers, while the default configuration can be to discard such packets.

When de-multiplexing the "next header" field in the IPv6 header under normal circumstances, the processing begins with the first extension header (the upper-layer header is processed directly in the absence of extension headers). The contents and semantics of each extension header determine whether to continue processing the next header. Therefore, it is very important that extension headers be processed according to the sequence in which they appear in the data packet. The recipient is unable to search for a particular extension header in the data packet to process it before all the headers until they are processed.

The information carried in the "hop-by-hop option" header must be examined and processed by each node on the data packet's delivery path, including the source and destination nodes. However, to improve forwarding efficiency, some high-performance routers may ignore the "hop-by-hop option" header or place data packets containing such headers in a lower-priority queue for processing. If the "hop-by-hop option" header is used, it must follow immediately after the IPv6 header. When the "next header" field in the IPv6 header has a value of 0, it means that there is a "hop-by-hop option" header following.

If the outcome of processing a header is for the node to process the next header while the node is unable to recognize the value in the "next header" field, then the node will discard the data packet and send a "Parameter Error" ICMP message to the source node. The ICMP code value of 1 is used in the message to indicate that the "next header"-type is not recognized. The ICMP pointer field contains the offset of the unrecognized field in the data packet. In the event that a node encounters a "next header" field with a value of 0 in any header (except the IPv6 header), the data packet should also be processed in the same way as described above by the node.

The length of each extension header should be an integral multiple of 8 (measured in bytes) to ensure that subsequent headers are also aligned by 8 bytes. Multi-byte fields within each extension header are aligned according to their natural demarcation.

The complete implementation of IPv6 includes the following extension headers:

- Hop-by-hop options;
- Routing;
- Fragmentation;
- Destination Options;
- Authentication (See Note 1 and Note 3);
- Encapsulating security payload (See Note 2 and Note 3).

Note 1: The authentication header is used to provide connectionless integrity and initial data authentication for IP datagrams, as well as to prevent retransmission attacks.

Note 2: The encapsulating security payload header is used to provide confidentiality, initial data authentication, connectionless integrity, protection against retransmission attacks, and the confidentiality of restricted data flow.

Note 3: Both the authentication and encapsulating security payload headers can be used together or through tunneling nestification. They can provide security services between hosts, between security network management systems, or between hosts and security network management systems. Additionally, encapsulating security payload headers can also provide confidentiality services such as encryption. The main difference between the two is the range of coverage. In addition, if the IP header is not encapsulated by the encapsulating security payload (in tunneling mode), the encapsulation security payload header will not offer protection to any field in the IP header.

(1) The order of extension headers
When a data packet contains multiple extension headers, they should be arranged in the following order:

- IPv6 header;
- Hop-by-hop option header;
- Destination option header (see Note 1);
- Routing header;
- Fragmentation header;
- Authentication header (see Note 2);
- Encapsulating security payload header (see Note 2);
- Destination option header (see Note 3);
- Upper-layer header.

Note 1: These options are processed at the first destination listed in the IPv6 "destination address" field, and at the subsequent destinations listed in the routing header.

Note 2: Additional suggestions on the relative order between the authentication header and encapsulating security payload header are given in IETF RFC 1827.

Note 3: These options are only processed at the final destination of the data packet.

Each type of extension header can appear once at most. If there are several extension headers of the same type, they should be arranged in sequence. The only exception is the destination option header, which can appear twice – once in the routing header, and again in the upper-layer header.

If the upper-layer header is another IPv6 header (i.e. IPv6 encapsulated in another IPv6 through tunneling), then its own extension headers follow immediately after. These extension headers are also to be arranged in the order specified above.

Should there be a need to define other extension headers, their position with respect to the order of the headers listed above needs to be specified.

IPv6 nodes must accept and process all extension headers in the same data packet regardless of their order and the number of times they appear. The only strict requirement is for the "hop-by-hop option" header to follow immediately after the IPv6 header. However, it is highly recommended that the sender of a data packet strictly adheres to the order suggested above unless subsequent specifications supersede it.

(2) Options

Among the extension headers introduced earlier, both the "hop-by-hop option" header and the "destination option" header can carry a varying number of options. These options are TLV-encoded, as shown in Figure 3-15.

Option Type: An unsigned 8-bit integer that describes the type of option.

Opt Data Len: An unsigned 8-bit integer (measured in bytes) representing the length of the option data.

Option Type	Opt Data Len	Option Data

Figure 3-15: Format of an IPv6 extension header option

Option Data: This is a variable length field that contains "Option Type" data.

When a recipient is processing a header, it must deal with the options according to the order in which they appear in the header. For example, searching for an option in the header and processing it before the options listed in front of it is disallowed.

In internal coding, the two most significant bits of the "Option Type" field indicate the actions that must be taken when the IPv6 node is unable to identify the type of option:

00 – Skip this option and continue to process the header;

01 – Discard this data packet;

10 – Discard this data packet and send an ICMP data packet to the source address of the said data packet to indicate the type of unrecognized option, regardless of whether the destination address of the said packet is a multicast address.

11 – Discard this packet, and send an "Parameter Error" ICMP message to the source address of the said data packet if the destination address is not a multicast address. The code value is 2 and the pointer field points to an unrecognized option type.

The third bit in the "Option Type" field indicates whether the data for this option can change the routing for the data packet to reach its final destination. When there is an authentication header in a data packet, for any option whose data can change the routing, the entire data field of the option must be treated as zero when calculating or verifying the authentication value of the data packet.

0 – Option data does not change the routing;

1 – Option data may change the routing.

The three most significant bits mentioned above should be considered as part of the "Option Type", instead of being independent from it. That means the entire 8-bit "Option Type" should be used to identify a special option instead of just the least significant 5-bit "Option Type".

Both the "hop-by-hop option" header and the "destination option" header use the same "Option Type" numbered space. However, a special option may be forcibly used for only one of the two headers.

Some options have unique alignment requirements to ensure that the multi-byte value in the "Option Data" field conforms to the natural demarcation. The alignment requirement for an option is represented by xn + y. This means that the "Option Type" must appear in a position that is an integral multiple of x bytes plus y bytes from the beginning of the extension header.

E.g.:

2n   Any 2-byte integral multiple offset from the start of the extension header.
8n + 2   Any 8-byte integral multiple plus two bytes offset from the start of the
          extension header.

There are two types of padding options available. They are used to arrange subsequent options when needed, so that the length of the extension header is an integral multiple of 8 bytes. All IPv6 nodes must be able to recognize these padding options.

(3) The "hop-by-hop option" header

The "hop-by-hop option" header is used to carry information that must be examined by each node on the data packet's delivery path. If the value of the "next header" of the IPv6 header is zero, then the next header of the IPv6 header is the "hop-by-hop option" header.

The format of the "hop-by-hop option" header is shown in Figure 3-16.

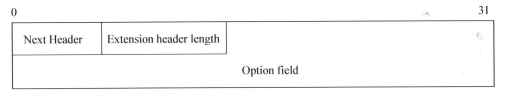

Figure 3-16: Format of a "hop-by-hop option" header

Next Header: This field is 8 bits long and defines the type of header that immediately follows the "hop-by-hop option" header. This field takes different values corresponding to different types of extension headers. For the correspondence between values and extension header types, please refer to the IANA protocol numbers and assigned services website.

Extension Header Length (Hdr Ext Len): This field is an 8-bit unsigned integer that is measured in 8-byte units. It represents the length of the "hop-by-hop option" header, excluding the first 8 bytes.

Options: This field varies in length, and its length should ensure that the length of the entire "hop-by-hop option" header is an integral multiple of 8 bytes. The field contains one or more TLV-encoded options.

Currently only "Pad1" and "PadN" options are specified for "hop-by-hop option" headers.

(4) Routing headers

A routing header is used in IPv6 source data packets to list one or more intermediate nodes that the data packet needs to pass through in its journey from the source address to the destination address. This feature is very similar to the loose source routing and record route options in IPv4. If a header has a value of 43 in its "next header" field, then the next header follows is the routing header.

The format of a routing header is shown in Figure 3-17

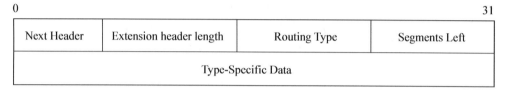

Figure 3-17: Format of a routing header

Next Header: This field is 8 bits long and defines the type of header that immediately follows the routing header. This field takes different values corresponding to different types of extension headers. For the correspondence between values and extension header types, refer to the "IANA protocol number and assigned services website".

Extension Header Length (Hdr Ext Len): This field is an 8-bit unsigned integer that is measured in 8-byte units. It represents the length of the routing header, excluding the first 8 bytes.

Routing Type: This field is 8 bits long and identifies the different types of routing headers.

Segments Left: This field is an 8-bit unsigned integer representing the number of routing segments remaining. That is the number of intermediate nodes that have been listed but not yet visited before arriving at the final destination node.

Type-Specific Data: This field varies in length, and its format is determined by the "Routing Type". Its length should ensure that the length of the routing header is an 8-byte integral multiple.

While processing a received data packet, if a node encounters a routing header that contains a "Routing Type" value that is unrecognizable, then the node takes measures based on the value of the "Segments Left" field. The specific method is described below:

- If the value of "Segments Left" is zero, the node ignores the routing header and proceeds to process the next header in the data packet, whose type is determined by the value of the "Next Header" field in the routing header.
- If the value of "Segments Left" is not zero, the node must discard the data packet and an "Parameter Error" ICMP message (Code 0) should be sent to the source node. The ICMP pointer points to the unrecognized "Routing Type".

After processing the routing header of a received data packet, if an intermediate node decides to forward the said data packet to a link with an MTU smaller than the length of the data packet, the node must discard the data packet and send an ICMP error message saying "data packet too big" to the source address.

(5) Fragmentation headers

IPv6 source nodes use fragmentation headers to send data packets that are longer than the MTU of the path. If the "Next Header" field of a header has a value of 44, then the header that follows immediately after is the fragmentation header.

The format of a fragmentation header is shown in Figure 3-18.

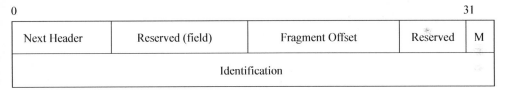

Figure 3-18: Format of the fragmentation header

Next Header: This field is 8 bits long, and defines the type of header that immediately follows the routing header. This field takes different values corresponding to different types of extension headers. For the correspondence between values and extension header types, please refer to the IANA protocol numbers and assigned services website.

Reserved (Reserved Field): This field is 8 bits long, with an initial value of zero at the time of transmission. This field is ignored by the recipient.

Fragment Offset: The field is 13 bits long, and is measured in 8-byte units. It specifies the data offset by the data behind the header relative to the starting position of the fragmentable part of the original data packet.

Reserved: This field is 2 bits long, with an initial value of zero at the time of transmission. This field is ignored by the recipient.

M Flag: This field is 1 bit long. M=1 means more segments to come, while M=0 means that the current segment is the last.

Identification: This field is 32 bits long. A detailed description is given below.

In order to transmit a data packet larger than the MTU of the path from the source node to the destination node, the source node may fragment the data and then send each fragment as an independent data packet for reassembly by the recipient.

IPv6 nodes must not generate overlapping fragments when sending data packets that need to be fragmented. When a receiving node is carrying out reassembly, it has to discard all relevant data packets if an overlapping fragment is found.

The source node generates an identification value for each data packet to be fragmented. The identification value must be different from that of any other data packet sent recently (see Note below) from the same source address to the same destination address. If there is a routing header, then the destination address mentioned above refers to the final destination address.

Note: "Recently" refers to a point within the maximum possible lifespan of a data packet. This includes the transmission time from source to destination and the waiting time needed for the reassembly of all the fragments of a data packet. However, the source node does not need to know the maximum lifespan of the data packet. The requirement can be assumed to have been met by using the identification value as a 32-bit cyclic counter whose count increases by one whenever a data packet is fragmented. This information is then added to the identification field. This is used to ensure the uniqueness of the identification. Each IPv6 implementation can decide between being configured with single or multiple node counters i.e. one counter for each possible source address, or one counter for each pair of source and destination addresses.

The original data packet refers to the original, unfragmented data packet. It consists of the two parts shown below.

The form of the original data packet is shown in Figure 3-19.

Non-fragmentable part	Fragmentable part

Figure 3-19: Form of the original data packet

The non-fragmentable part includes the IPv6 header and all extension headers to be processed by nodes on the path to the destination, such as all headers including routing headers or "hop-by-hop option" headers (if any), or no extension header at all.

The fragmentable part includes the rest of the data packet, with extension headers, upper-layer headers, and data that can only be processed by the destination node.

The fragmentable part of the original data packet is divided into fragments. The length of each fragment is an integral multiple of 8 bytes except for the last fragment. Each fragmented data packet is transmitted according to the manner shown in Figure 3-20.

Original data packet

Non-fragmentable part	Fragment 1	Fragment 2	......	Fragment n

Fragmented Data Packet

Non-fragmentable part	Fragmentation Header	Fragment 1

Non-fragmentable part	Fragmentation Header	Fragment 2

Non-fragmentable part	Fragmentation Header	Fragment n

Figure 3-20: Example of data packet fragmentation

The composition of each fragmented data packet is as follows:

- Non-fragmentable parts of the source data packet. The payload length of the source data packet's IPv6 header is changed to the length of the fragmented data packet (excluding the length of the IPv6 header itself). In addition, the value of the "Next Header" field of the last header in the non-fragmentable part is changed to 44.
- Fragmentation header
  - **Next Header**: Indicates the first header of the fragmentable part of the original data packet.
  - **Segments Left**: Measured in 8-byte units, this indicates the offset of a fragment relative to the starting position of the fragmentable part of the original data packet. The value of the "Segments Left" field of the first fragment is zero.
  - **M Flag**: M = 0 means the last fragment, while M = 1 means that the fragment is not the last.
  - **Identification**: Used to identify the original data packet.

- Fragment
  The length of the fragmented data packet must not exceed the MTU of the path to the data packet's destination.

At the destination node, the fragmented data packets are reassembled to restore the original, unfragmented data packet.

The reassembled original data packet is shown in Figure 3-21.

Non-fragmentable part	Fragmentable part

Figure 3-21: Reassembled original data packet

The principles for reassembly are as follows:

The original data packet can only be reassembled from fragmented data packets with the same source address, destination address, and fragment identification.

The non-fragmentable part of the reassembled data packet includes all headers except the fragmentation header of the first fragmented data packet. There are two main changes to this part:

- The value of the "Next Header" field of the last header in the non-fragmentable part is that of the "Next Header" field in the fragmentation header of the first fragment;
- The payload length of the reassembled data packet is calculated using the length of the non-fragmentable part, and the length and offset of the last fragment. The formula for calculating the length of the reassembled data packet is:

$$PL.orig = PL.first-FL.first-8 + (8 * FO.last) + FL.last$$

Where:

PL.orig = Payload length value of the reassembled data packet;

PL.first = Payload length value of the first fragmented data packet;

FL.first = The length of the fragment after the fragmentation header of the first fragmented data packet;

FO.last = The "Segments Left" value in the fragmentation header of the last fragmented data packet;

FL.last = Length of the fragment after the fragmentation header in the last fragmented data packet.

The fragmentable part of the reassembled data packet is made up of the fragment after the fragmentation header of every fragmented data packet.

The length of each fragment is equal to the data packet payload length minus the length of the header between the IPv6 header and the fragment; the relative position of each fragment in the "fragmentable part" of the data packet is calculated using the value of the "Segments Left".

The fragmentation header does not appear in the final reassembled data packet.

The following errors may occur in the reassembly of fragmented data packets. If all fragments of a data packet that are to be reassembled do not arrive within 60 seconds after the receipt of the first fragment, the reassembly is discontinued and all received fragments are to be discarded. In such a scenario, if the first fragmented data packet has been received, an "Timeout-Fragment Reassembly Timeout" ICMP message has to be sent to the source address of that fragmented data packet.

If the fragment length obtained from the "Payload Length" field of a fragmented data packet is not an 8-byte integral multiple and the M Flag of this segment is 1, then this fragment must be discarded, and an "Parameter Error" ICMP message (Code 0) should be sent to the source address. The ICMP pointer points to the "Payload Length" field in the fragmented data packet.

If the length and offset of a fragment causes the "Payload Length" of the reassembled data packet to exceed 65535 bytes, then the fragment must be discarded and an "Parameter Error" ICMP message (Code 0) should be sent to the source address. The ICMP pointer points to the "Segments Left" field of the fragmented data packet.

Although the following scenarios are not desirable during reassembly, they are not considered to be errors when they occur.

The number and content of the headers before the fragmentation header of different fragments belonging to the same data packet can differ. Regardless of their type, all headers before the fragmentation header in each fragment data packet have to be processed upon arrival of the data packet before it enters the reassembly queue. Only the headers of fragmented data packets with an offset value of zero are retained in the reassembled data packet.

The value of the "Next Header" field in the fragmentation header of different fragments arising from the same original data packet can be different. The reason is that only the corresponding value of fragmented data packets with an offset of zero are useful for reassembly.

(6) Destination option headers

The destination option header carries optional information that only needs to be processed by the destination node. When the "Next Header" field of any header has a value of 60, it means that the header immediately after is the destination option header. Its format is shown in Figure 3-22.

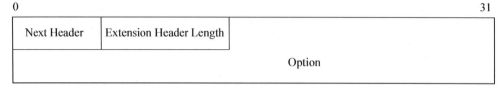

Figure 3-22: Destination option header format

Next Header: This field is 8 bits long, and defines the type of header that immediately follows the "destination option" header. This field takes different values corresponding to different types of extension headers. For the correspondence between values and extension header types, please refer to the IANA protocol numbers and assigned services website.

Extension Header Length (Hdr Ext Len): This field is an 8-bit unsigned integer that is measured in 8-byte units. It represents the length of the "destination option" header, excluding the first 8 bytes.

Options: This field varies in length, and its length should ensure that the length of the entire "destination option" header is an integral multiple of 8 bytes. The field contains one or more TLV-encoded options.

Currently only "Pad1" and "PadN" options are specified for "destination option" headers.

Note that there are two ways to encode optional destination information in an IPv6 data packet – either as an option in the "destination option" header, or as a separate extension header. Examples of the latter include both the fragmentation and authentication headers. The method chosen depends on the course of action the destination node should take if it does not recognize the optional information.

- If it is a desired action for the destination node to discard the data packet and send an "Unrecognized Type" ICMP message to the source address of the data packet (provided that the destination address is not a multicast address), then both ways are feasible for the encoding of information. Specifically, for the "destination option" header method, the two most significant bits in the "Option Type" field must have a value of 11. Some criteria for choosing between the two methods include the number of bytes used (fewer is better), ease of arrangement, and efficiency of semantic parsing.

- If other actions are desired, then the information needs to be encoded as an option in the "destination option" header with the value of the two most significant bits in the "Option Type" field as 00, 01, or 10.

(7) No next header

When the value of the "Next Header" field in the IPv6 header or any extension header is 59, it means that there is no more data after this header. If the "Payload Length" field of the IPv6 header indicates that there are additional bytes following the header, and these bytes follow a header with a value of 59, then these bytes must be ignored. However, if the data packet is to be forwarded, then it is allowed to continue its journey unchanged.

3.4 IPv6 Basic Protocol

3.4.1 IPv6 Neighbor Discovery Mechanism

Nodes (hosts and routers) make use of neighbor discovery to determine the link-layer address of neighbors on the connected link, and quickly delete invalid cache values. Hosts also use neighbor discovery to find neighboring routers for the forwarding of packets. In addition, nodes use the neighbor discovery mechanism to proactively track which neighbors are reachable or unreachable, and to detect link-layer addresses that have changed. When a router malfunctions or a path to the router is no longer passable, the host actively looks for another router or path that is still working.

The IPv6 neighbor discovery protocol corresponds directly to the ARP, ICMP router discovery, and ICMP redirect in IPv4. However, there is no equivalent neighbor unreachability detection mechanism and protocol in IPv4.

Neighbor discovery supports the following types of links: point-to-point, multicast, NBMA, shared media, variable MTU, and asymmetric reachability.

1) Functions of the neighbor discovery mechanism

The neighbor discovery mechanism has the following functions:

- **Router discovery**: How a host locates the routers on the connected link;
- **Prefix discovery**: How a host discovers a set of address prefixes, and which of the destinations on the connected link defined by the set of prefixes are on-link;
- **Parameter discovery**: How a node obtains the link parameters (such as the MTU of the link) or the network parameters (such as the hop limit) of the sending interface;

- **Automatic address configuration**: How a node automatically configures the address of the interface;
- **Address Resolution**: How a node determines the link-layer address of an on-link destination such as a neighbor when given the destination IP address;
- **Next-hop determination**: An algorithm that maps a destination IP address to the IP address of a neighbor. The next hop can be a router or a destination;
- **Neighbor unreachability detection**: How a node determines that a neighbor is unreachable. If the neighbor is a router, the default router can be used. However, if the neighbor is both a router and a host, address resolution needs to be carried out again;
- **Duplicate address detection**: How a node determines that an address to be used is not being used by another node;
- **Redirection**: How the router informs the host of the optimal next-hop to the destination.

2) Neighbor discovery to define the ICMP packet type

Five different ICMP packet types are defined based on neighbor discovery. These are the router solicitation and router advertisement messages, neighbor solicitation and neighbor advertisement messages, and the redirect messages.

(1) Router solicitation
When the interface is working, the host sends a router solicitation message immediately, requesting that the router generates a router advertisement message without having to wait for the next scheduled time.

(2) Router advertisement
The router periodically advertises its presence and its configured links and network parameters, or responds to router solicitation messages. The router advertisement message contains the prefix of on-link determination, the prefix of the address configuration, and the hop limit value.

Router discovery is part of the basic protocol, and the host does not have to probe the routing protocol.

On multicast links, each router periodically multicasts route advertisement packets to advertise its availability. A host receives router advertisements from all routers, and establishes a default router list. Routers frequently generate router advertisements to let a host know their availability within the next few minutes. This also allows the host to carry out fault detection using neighbor unreachability detection algorithms.

The router advertisement contains a list of prefixes used for on-link determination and/or address configuration, as well as prefix flag-bits for indicating the use of a particular prefix. The host uses the on-link prefixes in the advertisement to set up and maintain the list. The list is used to determine if the destination of the packet is on-link or outside the router. Even if the destination is not included in the on-link prefix in the advertisement, a connection to the destination is still possible. In this case, the router sends a redirect message to notify the sender that the destination is a neighbor.

The router advertisement allows the router to inform the host how to perform address configuration. For example, the router can direct the host to configure using a state or a stateless address.

The router advertisement message contains network parameters such as the hop limit used by the sending interface of the host, and optional link parameters such as the link MTU. This helps to centralize the management of these important parameters that are set on the router so that they are automatically sent to all connected hosts.

The router advertisement contains the link-layer address and does not need any additional packet exchange to resolve the link-layer address of the router. With the link prefix, no separate mechanism to configure the mask is required.

The router advertises the MTU used by the host on the link to ensure that all nodes on the link use the same value of MTU.

(3) Neighbor solicitation

The node sends a neighbor solicitation message to determine the link-layer address of the neighbor or to verify that the neighbor's link-layer address in the cache is still reachable. The neighbor solicitation message can also be used for duplicate address detection.

The node completes the address resolution by multicasting the neighbor solicitation message, and the neighbor solicitation message requires the target node to send back its link-layer address. The neighbor solicitation message is sent by multicasting to the solicited node's multicast address of the target node. The target returns its link-layer address by unicasting a neighbor advertisement message. The originator's neighbor solicitation message includes its link-layer address.

Neighbor solicitation messages can also be used to determine if multiple nodes are assigned the same unicast address. Neighbor solicitation messages for duplicate address detection are specified in the automatic address configuration.

(4) Neighbor advertisement

This is a response to the neighbor solicitation message. A node can also send an unsolicited neighbor advertisement to indicate changes to the link-layer address.

Neighbor unreachability detection can detect any fault with the neighbor, or in the neighbor's forward path. This requires confirmation that the packet sent to a neighbor has reached said neighbor and is being processed by the IP layer. Neighbor unreachability detection is carried out using two methods. The first is that the upper protocol provides confirmation that a link is being processed, i.e. the data sent previously has been sent correctly (e.g. through a new confirmation received recently). The other is for the node to send a unicast neighbor solicitation message. A neighbor advertisement message from the solicited node is confirmation of the reachability of the next hop. To reduce unnecessary network traffic, probe messages are sent only to neighbors.

In the event of link or node failure such as a malfunctioning router or changes to the link-layer address, neighbor unreachability detection plays a part in ensuring that a packet is delivered. For example, an invalid ARP cache can cause a mobile node to be left out of a connection even though it has not lost any connection. Unlike the ARP, neighbor discovery detects half-link failures, and the sending of traffic to neighbors that have lost their connectivity in both directions is avoided.

Router advertisement messages do not contain priority fields and do not have to deal with routers of differing stability. Neighbor unreachability detection detects routers that are no longer working, and switch to routers that are working.

(5) Redirect

This is used by the router to inform the host of the optimal next hop to reach the destination.

The redirect contains the link-layer address of the new first hop. Redirection is unnecessary for addresses that are resolved separately.

Multiple prefixes can be related to the same link. By default, the host obtains all the on-link prefixes from the router advertisement. However, the router can configure some or all of the prefixes in the router advertisement to be ignored. In this case, the host considers the destination to be non-connected. Traffic is then sent to the router before it issues the appropriate redirect.

Receipt of the IPv6 redirect message implies that the next hop is connected. In IPv4, the network mask of the link indicates that the next hop is not connected, and the host ignores the redirect. The IPv6 redirect mechanism is similar to the redirect mechanism of shared media. In non-broadcast and shared media links, it is not possible for a node to know all the prefixes at the connection destination.

In addition to dealing with the above mentioned general problems, neighbor discovery can also handle the following situations.

- **Changes to the link-layer address**: If a node learns that its link-layer address has changed, it will multicast neighbor advertisement packets to all nodes for them to quickly update the invalid link-layer addresses in their cache. Sending unsolicited advertisement messages can only improve reliability when the link is unreliable. The neighbor unreachability detection algorithm ensures that all nodes have a reliable method to find the new addresses, but this may increase the delay.

- **Ingress load-balancing**: Load-balancing is required when receiving data packets from multiple network interfaces on the same link. This node is assigned multiple link-layer addresses on the same interface. For example, a single network driver can designate multiple network interface cards as a logical interface with multiple link-layer addresses. Load-balancing allows the router to omit the link-layer address of the source from the router advertisement packet, and forces the neighbor to use the neighbor solicitation message in order to obtain the link-layer address of the router. The sender of the solicitation message is different, and the neighbor advertisement message that is returned contains different link-layer addresses.

- **Anycast address**: An anycast address is used to identify a group of nodes that provide the same service. Multiple nodes on the same link can be configured to identify the same anycast address. The node is used to receive multiple neighbor advertisements and neighbor discovery for the same destination to handle anycast. All anycast address advertisements are marked as non-override advertisements so that specific rules can be employed to determine which advertisement to use.

- **Proxy advertisement**: If the destination address is unable to respond to neighbor solicitations, the router that is willing to receive the packets on behalf of the target address will issue a non-override neighbor advertisement. There is currently no specification for the use of proxies. Proxy advertisements can be used to handle mobile nodes that have left off-link, but not as a general mechanism for dealing with nodes.

Using a link-local address to uniquely identify a router (for router advertisement and redirect messages), the host can maintain contact with the router even while the site is renumbered, and a new global prefix is used.

The hop limit in neighbor discovery is 255 to prevent non-connected senders from intentionally sending neighbor discovery messages. In IPv4, senders send both redirect messages and router advertisement messages.

3.4.2 ICMPv6 protocol

ICMP is defined for IPv4. For it to be used by IPv6, many changes have been made. A value of 58 in the "next header" field of IPv6 indicates an ICMPv6 message. ICMPv6 is used by IPv6 nodes to report errors that occur during packet processing and to perform other network interconnection functions, such as using the ICMPv6 "PING" for troubleshooting. ICMPv6 is an integral part of IPv6 and one of the underlying protocols. It is regulated by IETF RFC 4443. All messages and behaviors specified in the RFC must be performed unconditionally by each IPv6 node.

There is an IPv6 header and several IPv6 extension headers before each ICMPv6 message. The ICMPv6 header has a value of 58 in the "next header" field.

Figure 3-23 shows the general format of an ICMPv6 message.

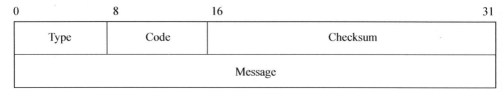

0	8	16	31
Type	Code	Checksum	
Message			

Figure 3-23: General format of an ICMPv6 message

The fields are described as follows:

- **Type**: Indicates the type of message. The value of this field determines the rest of the data format.
- **Code**: Depends on the type of message. It is used to generate additional layers for message granularity.
- **Checksum**: Used to detect data errors in the ICMPv6 message and IPv6 header.

ICMPv6 messages fall into two categories—error messages and instructional messages. Error messages are identified by the binary value of 0 for the most significant bit in the "Type" field.

3.5 IPv6 routing mechanism

IPv6 adopts a routing mechanism that is similar to that of IPv4. The IPv6 routing system consists of two parts—internal routing and external routing.

Autonomous systems (AS) describe the transition of a group of routers from internal routing to external routing. IP data packets typically travel through the routers of two or more AS to reach their destination. To support such forwarding, AS need to provide topological information to each other. The interior gateway protocols are used to distribute routing information within the AS, i.e. internal routes. The exterior gateway protocols are for the exchange of routing information between AS, i.e. external routes.

3.5.1 Interior gateway protocols

1) Definition

Interior gateway protocols (IGP) are used to distribute routing information among specific routers inside the AS. The implementation of a specific IGP algorithm is relatively independent, but the following functions must be realized:

- It should be able to quickly reflect changes in the internal topology of the AS;
- It must provide a mechanism to prevent constant route-updating caused by circuit oscillations;
- It must provide fast convergence to loop-free routing;
- It must minimize bandwidth used;
- It must provide equivalent routing for load sharing;
- It must provide a certified route-updating method.

Currently, there are three types of IGPs specified in IPv6: RIPng, OSPFv3, and IS-ISv6.

2) Shortest Path First Protocol version 3 (OSPFv3)

This is a shortest path first (SPF) protocol based on the link-state algorithm. It is based on the Dijkstra shortest path algorithm. In SPF-based systems, each router obtains a complete topological database through a process called a flooding algorithm. Flooding ensures that information is transmitted reliably. Each router that runs the SPF algorithm establishes an IP routing table on the data path.

In IPv4 networks, version 2 of the OSPF protocol is used. This is often referred to as OSPFv2. In order to support IPv6 networks, the IETF has revised OSPFv2 to come up with OSPFv3.

The IPv6 OSPF protocol retains most of the IPv4 algorithms. From IPv4 to IPv6, the basic OSPF mechanism remains unchanged. Both IPv6 and IPv4 contain a link-state database. A link-state advertisement (LSA) message is stored in the link-state database. The information stored in neighboring routers is kept in sync. Initial database synchronization is achieved through the database exchange process.

This process includes the exchange of database description packets, link-state request packets, and link-state update packets. Subsequent database synchronization is maintained by flooding, which makes use of link-state update packets and link-state acknowledgement packets to complete the synchronization. In both broadcast and non-broadcast multiple-access (NBMA) networks, the OSPF Hello packet is used in both IPv6 and IPv4 to discover and maintain neighbor relationships, as well as to elect and back up designated routers.

IPv6 and IPv4 are also consistent in other aspects, for example, the basic ideas behind determining which neighbors becomes adjacent, inter-domain routing, introducing external information such as LSA outside an AS, and the calculation of different routes.

The following OSPF features in IPv4 are unchanged in IPv6:

- Both IPv4 and IPv6 use the same packet types, i.e. Hello packets, database description packets, link-state request packets, link-state update packets, and link-state acknowledgement packets. In certain cases (such as for the Hello packet), there are some changes to the packet format, but the functionality remains unchanged.
- As OSPF operates directly on the IPv6 network layer, OSPF in IPv6 requires the IPv6 protocol stack, but the OSPF system requirements are unchanged.
- Discovery and maintenance of neighbor relationships, as well as the selection and establishment of adjacencies remain unchanged. This includes the election and back up of designated routers in both broadcast and NBMA networks.
- OSPF supports the same link type (or interface type), i.e. point-to-point, broadcast, NBMA, point-to-multipoint, and virtual links.
- Interface state machines (including the list of OSPF interface states and events) and the election algorithm for both designated routers and back-up designated routers are unchanged.
- Neighbor state machines (including the list of OSPF neighbor states and events) remain unchanged.
- Aging of the link-state database, and the updating of LSA in the routing domain through the unexpired aging process are unchanged.

(1) Differences in OSPF functions between IPv4 and IPv6
The semantic changes and increase in address space of the IPv6 protocol mean that there are a number of differences between the OSPF protocols in IPv6 and IPv4.

• The protocol is handled on the link instead of on the subnet

In IPv6, the term "link" is used to indicate a communication device or medium through which nodes communicate at the link level. "Interfaces" are connected to links, and multiple IP subnets can be assigned to a link. Two nodes that are not on the same IP subnet can communicate directly on a single link.

Therefore, OSPF in IPv6 operates on every link, while OSPF in IPv4 operates on every IP subnet. The terms "network" and "subnet" in the OSPF specifications of IPv4 are replaced by "link", and the OSPF interface is connected to the link rather than to the IP subnet.

• Deleting address semantics

In OSPF based on IPv6, address semantics have been deleted from the OSPF protocol packets and the major LSA types.

· Other than the LSA payload carried in the link-state update packet, the IPv6 address does not appear in the OSPF packet;

· The router LSA and network LSA no longer include the network address. Instead, the topological information is represented simply;

· As in IPv4, both the OSPF router ID and LSA link-state ID are 32 bits long, but they are no longer assigned IPv6 addresses;

· Neighbor routers are identified by the router ID, and no longer use their IP addresses on broadcast and NBMA networks.

• Increased flooding range

The flooding range of an LSA is indicated in the LS-type field. There are three flooding ranges:

· Link-local: LSA flooding only occurs on the local link. A link LSA uses this range.

· Domain: LSA flooding only occurs for an OSPF domain. Router LSA, network LSA, inter-domain prefix LSA, inter-domain router LSA, and intra-domain prefix LSA use this range.

· Autonomous domains: LSA flooding occurs in the routing domain. AS-external LSA use this range.

• Each link supports multiple instances

OSPF supports the running of multiple OSPF protocol instances on a single link. For example, at a NAP (network access point) shared by several operators, all operators have one or more physical network segments (e.g. links) in common. In this case, the operator can operate an independent OSPF routing domain.

- Usage of link-local addresses

IPv6 link-local addresses are used on a single link for neighbor discovery and auto-configuration. IPv6 routers do not forward IPv6 data packets that contain link-local source addresses. The range of IPv6 addresses assigned to link-local unicast addresses is FF80::/10.

Each router is assigned a link-local unicast address on a connected physical segment. Other than all OSPF interfaces on virtual links, all link-local addresses associated with interfaces can be used as the source address to send OSPF packets. The router is able to obtain the link-local address of all other routers connected to the link, and these addresses can be used as the next-hop information for packet forwarding.

On a virtual link, OSPF protocol packets must use the IP address of a global or local site as the source address.

Link-local addresses can appear in OSPF link LSA but not in other types of LSA. In addition, link-local addresses cannot be advertised in inter-domain prefix LSA, AS external LSA, or intra-domain prefix LSA.

- Changes of verification

The authentication type and authentication field are deleted from the OSPF header in IPv6. No authentication-related fields will now appear in the OSPF domain and interface structure.

In IPv6, the OSPF uses the IP authentication header and IP-encapsulating security payload to ensure the integrity and confidentiality of data.

- Changes to the packet format

IPv6 OSPF operates directly on IPv6. The OSPF packet header does not contain the address semantics. They are included in the different LSA types instead. As such, the OSPF in IPv6 has nothing to do with network protocols.

Changes to the OSPF packet format are as follows:

- The OSPF version number has increased from 2 to 3;
- The option field for both the Hello packet and the database description packet are extended to 24 bits;
- The authentication and authentication-type fields have been removed from the OSPF packet header;

· The Hello packet does not include address semantics. Instead, it contains an interface ID that identifies the link interface and is assigned to the originating router. If the router becomes the DR on the link, the interface ID is the link-state ID of the network LSA;

· In order to process the router LSA during SPF calculation, two option bits are added to the option field: the R-bit and V6-bit. If the R-bit is set to 0, the OSPF speaker can distribute topological OSPF information without forwarding traffic, such as when a multi-homed host needs to participate in the routing protocol. If the V6-bit is set to 0, the OSPF speaker can distribute topological OSPF information without forwarding IPv6 datagrams. If the R-bit is set to 1 and the V6-bit is set to 0, then IPv6 data packets are not forwarded. Instead, data packets of another protocol are forwarded;

· The OSPF packet header contains an "instance ID" that allows the running of multiple OSPF protocol instances on a single link.

• Changes to the LSA format

All address semantics are omitted from the LSA header, router LSA, and network LSA. The two LSAs describe the topology of the routing domain in a manner independent of the network protocol. In addition, new LSAs for the distribution of IPv6 address information, and data needed for next-hop resolution have been added.

The specific changes of the LSA format are as follows:

· The option field is omitted from the LSA header. Instead, the field is extended to 24 bits and added to the main body of the router LSA, network LSA, inter-domain router LSA, and link LSA;

· The LSA-type field is extended to 16 bits. The first three bits are used to decode the flooding range and handle unknown LSA types;

· In LSA, an address is represented by the prefix followed by the prefix length instead of the address followed by the mask. The default length of a routing prefix is zero;

· Both router and network LSAs do not contain address information. Therefore, they are unrelated to the network protocol;

· Router interface information can be diffused through multiple router LSAs. When performing SPF calculations, the recipient must connect all router LSAs generated by a given router;

· Link LSAs have a local link flooding range and will not be flooded outside the link. Link LSAs have three functions: 1) to provide the link-local address of the router. This address reaches all routers connected to the link; 2) to inform other routers and links about the connection between the routers and the links in the IPv6 prefix list; 3) to allow the router to maintain a collection of network LSA-related option fields, where the network LSAs are generated by the links;

· The third type of summary LSA has been renamed the "Intra-domain prefix LSA", and the fourth type renamed the "Intra-domain router LSA";

· The link state ID of the inter-domain prefix LSA, intra-domain router LSA, and AS-external LSA is only used to identify a single link-state database and does not include address semantics. All addresses or router IDs are no longer represented by the link state ID. Instead, they reside in the body of the LSA;

· The link state ID of network LSAs and link LSAs are the interface IDs of the initiating routers on the link. Therefore, there is no restriction on the size of network LSAs and link LSAs. All routers on the connected links must be listed in the network LSA. The link LSA must contain the address of all the routers on the link;

· The intra-domain prefix LSA contains all the information about IPv6 prefixes. In IPv4, this information is found in the router LSA and network LSA;

· The optional address of the AS-external LSA include forwarding addresses and external-routing labels. In addition, AS-external LSAs can refer other LSAs and route attributes outside the scope of the OSPF protocol, for example, when using BGP (border gateway protocol) path attributes and external routes.

• Handling of unknown LSA types

In IPv6, an unknown LSA type can be thought of as having a local link flooding range, or as a known LSA type for storage and flooding. In IPv4 OSPF, unknown LSA types are simply discarded.

• Support for stub areas

As with OSPF in IPv4, the purpose of having a stub area is to reduce the size of the link database and routing table of routers in the domain. As such, routers can use fewer resources to handle larger OSPF routing domains. In IPv6, a stub area can only handle router LSAs, network LSAs, and intra-domain prefix LSAs. Unknown LSA types can also be stored and flooded as known LSA types but this is controlled. Otherwise, it may cause the link state database of the stub area to expand beyond what the routers are capable of processing.

Therefore, the following rules are in place for the stub areas. If the LSA has an area or local link flooding range, and the U-bit of the LSA is set to 0, then an unknown LSA type can be flooded in the stub area.

- Using router ID to identify neighbors

In IPv6 OSPF, neighboring routers on a given link are identified by their OSPF router ID. In IPv4 OSPF, both neighbors on the point-to-point network and neighbors on virtual links are identified by their router ID. On the other hand, neighbors on the broadcast, NBMA, and point-to-multipoint links are identified by their IPv4 interface addresses.

The router ID of 0.0.0.0 is reserved and cannot be used.

- Protocol data structure

In both IPv4 and IPv6, the main OSPF data structures are the same, such as area, interface, neighbor, link state database, and routing table.

The top level of the data structure in IPv6 remains similar to that in IPv4 with the following changes. All known LS types and LSAs with AS flooding ranges no longer belong to a specific area or link, but come under the top level of the data structure. The AS-external LSA is the only LSA defined in this specification that has a flooding range. Unknown LS types, LSAs with a U-bit set to 1 (unrecognized but requiring flooding), and AS flooding ranges also fall into the top level of the data structure.

The IPv6 area data structure contains all the area-defined elements in IPv4, such as area ID, area address range list, relevant router interfaces, router LSA list, network LSA list, summary LSA list, shortest path tree, traversal capability, external routing capability, and stub area default cost. In addition, all known LSA types with area flooding ranges are also included in IPv6 area data structure. Typically, these include the router LSA, network LSA, inter-domain prefix LSA, inter-domain router LSA, and intra-domain prefix LSA. Both unknown LS types and LSAs with the U-bit set to 1 (unrecognized but requiring flooding) and area ranges also fall under the area data structure. IPv6 routers implementing MOSPF also need to add the LSAs of group members to the area data structure. Type 7 LSAs belong to the data structure of the NSSA area.

In IPv6 OSPF, the interface connects the router to the link. In IPv6, the following IPv4 interface structures are modified:

- **Interface ID**: Each interface is assigned an interface ID that uniquely identifies the router's interface. For example, in some implementations, the IfIndex of MIB-II may be used as the interface ID. The interface ID may appear in the following packets: Hello packets sent from the interface, link-local LSAs initiated by the router to connected links, and router LSAs initiated by the router LSA to the relevant areas. If the router is selected as the designated router, the interface ID can also be used as the link state ID of the network LSA;

· **Instance ID**: Each interface is assigned an instance ID. The default value is 0 but if the link contains multiple independent communities of OSPF routers, a different value would have to be assigned for these links. For example, if there are two groups of routers in an Ethernet segment, they need to be separated. Since the instance ID of all Ethernet interfaces for the said router is assigned a value of 0, the instance ID of the first group is 0. Other Ethernet interfaces of the said router are then assigned an instance ID of 1. In this way, the two groups are kept apart in both the sending and receiving processes of OSPF;

· **LSA list with link-local range**: All LSAs with a link-local range that initiate or flood on the link belong to the structure of the interface connected to this link. This includes the link LSA set of the link;

· **LSA list with unknown LS types**: All LSAs with unknown LS types and U-bits set to 0 (LSAs are processed with link-local flooding ranges if unrecognized) are stored in the data structure of the interface that receives this LSA;

· **IP Interface address**: For IPv6, the IPv6 address of the OSPF source packet sent by the interface is almost always the link-local address. The only exception is the virtual link, in which the router's global IPv6 address is used as the IP interface address;

· **List of link prefixes**: A list of IPv6 prefixes that can be configured for the connected links before being advertised by a router in the link LSA, so that the designated router can advertise it in the inter-domain prefix LSAs.

In IPv6 OSPF, each router interface has a measure that indicates the cost of sending packets from the interface. In addition, in IPv6, OSPF relies on the IP authentication header and IP encapsulating security payload to ensure the integrity and authentication/confidentiality of routing exchanges. Therefore, the AuType and authentication key have nothing to do with the OSPF interface in IPv6.

Interface state, events, and state machines are the same as in IPv4. The election algorithm for designated routers and designated back-up routers is also consistent with that in IPv4.

In both IPv6 and IPv4, the neighbor structure performs the same functions, i.e. if two routers need to establish an adjacency, the structure gathers all the information needed to do so. The differences in neighbor structures between IPv6 and IPv4 lie in the following aspects:

· **Neighbor interface ID**: The interface ID advertised by the neighbor in its Hello packet must be recorded in the neighbor structure. When a router advertises a point-to-point link to its neighbors, or when advertising a link to a neighbor that has become a designated router in the network, the neighbor's interface ID must be contained in the router LSA of the router;

· **Neighbor IP address**: Other than in virtual links, a neighbor's IP address is the link-local address in IPv6;

· **Neighbor's designated routers**: The election of a neighbor's designated router is encoded using the router ID instead of the IP address;

· **Neighbor's designated back-up routers**: The election of a neighbor's designated back-up router is encoded using the router ID instead of the IP address.

Neighbor state and even, and neighbor state machines remain unchanged from IPv4, as is the type of adjacency to be established.

Please refer to IETF RFC 5340 for details of the implementation of the OSPFv3 protocol.

3) IS-ISv6—Intermediate System to Intermediate System Protocol for IPv6

IS-ISv6 is based on the link state (SPF) routing algorithm and possesses all the advantages of this protocol type.

The IS-IS protocol as defined by ISO/IEC 10589 is an extensible routing protocol that supports the transmission of IPv4, IPv6, and OSI routing information. IETF RFC 1195 specifies the transmission method of the routing information in IPv4 (IS-IS), while IETF RFC 5308 specifies the transmission method of the routing information in IPv6 (IS-ISv6).

The IS-ISv6 protocol is an extension of the IS-IS protocol. Its basic message formats and procedures for the processing and transmission of routing information are identical to those of IS-IS. The main difference between the two is the addition of the attributes required for the transmission of IPv6 routing, and address information. Two new types of TLV are expanded, i.e. the reachability TLV and the interface address TLV. These are used to disseminate IPv6 routing information in the routing domain.

(1) IPv6 Reachability TLV

The TLV type value for the IPv6 reachability TLV is 236 (0xEC).

Two types of reachability TLVs are defined in IETF RFC 1195. They are the Internal IP Reachability Information and the External IP Reachability Information. In IS-ISv6, an IPv6 reachability TLV is used with an external bit to provide the equivalent IPv6 routing functions.

The IPv6 reachability TLV contains the routing prefix and metric information. It includes a U-bit to indicate if the prefix is being issued down from a higher level; an X-bit to indicate if the prefix is being issued from another routing protocol; and optional Sub-TLVs for future expansion. Through specifying this data, the IPv6 reachability TLV describes the reachability of a network. The format of the data is shown in Figure 3-24.

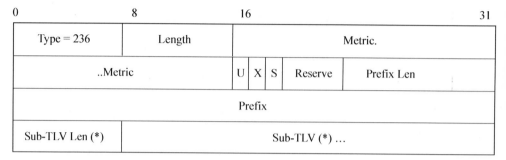

Figure 3-24: Format of the IPv6 reachability TLV

In the figure above, * denotes optional, U denotes the up/down bit, X denotes the external bit, and S denotes the sub-TLV bit.

The IPv6 reachability TLV can appear any number of times (or not at all) in the LSP. When a prefix is injected into IS-IS for the first time, the up/down bit is set to 0. If the prefix is reissued from a higher level to a lower level (e.g. from level 2 to level 1), the bit is set to 1 to indicate that the prefix is being issued from a higher to a lower level. If the prefix is reissued from one area to another area at the same level, the up/down bit is set to 1.

If the prefix is issued into IS-IS from another routing protocol, then the external bit is set to 1. This information is useful when issuing prefixes from IS-IS to other protocols.

If the Sub-TLV bit is set to 0, then the Sub-TLV is empty. Otherwise, if the bit is set to 1, then the byte following the prefix describes the length of the Sub-TLV, and the format includes the Sub-TLV section. The prefix is packed into the data structure. This means that only a few of the bytes that the prefix needed will appear in the data structure. The number of bytes can be calculated based on the prefix length as shown below:

Number of bytes of prefix = Integer of ([prefix length +7] / 8)

If an issued prefix has a metric greater than MAX_V6_PATH_METRIC (0xFE000000), the prefix is not considered during normal SPF calculations. Under this approach, it is permissible to issue a prefix for reasons other than establishing a normal IPv6 routing table.

If there is a sub-TLV, its format is the same as that of a normal TLV, as shown in Figure 3-25.

0	8	16	31
Type	Length	Metric (*) ...	

Figure 3-25: Format of the sub-TLV of an IPv6 reachability TLV

In the figure, * denotes optional, the Length field indicates how many bytes the Metric field has, and the value of the Length field can be zero.

(2) IPv6 interface address TLV
The TLV type value for the IPv6 interface address TLV is 232 (0xE8).

This TLV maps directly to the IP Interface Address TLV defined in IETF RFC 1195. To do this, the content needs to be adjusted. The 16-byte IPv6 interface address with the sequence number of 0 to 15 replaces the 4-byte IPv4 interface address with the sequence number 0 to 63. The data format is shown in Figure 3-26.

0	8	16	31
Type = 232	Length	Interface Address 1 (*) ..	
..Interface Address 1 (*) ..			
..Interface Address 1 (*) ..			
..Interface Address 1 (*) ..			
Interface Address 1 (*) ..		Interface Address 2 (*) ..	

Figure 3-26: IPv6 interface address TLV

Further restrictions on the syntax of the TLV are imposed according to whether the TLV is issued. For a Hello PDU, the interface address TLV must only contain the IPv6 link-local address assigned to the interface that sent the Hello message. For LSPs, the interface address TLV must only contain the non-link-local IPv6 address assigned to the IS.

(3) IPv6 NLPID
The value of IPv6 NLPID is 142 (0x8E).

According to the specifications of IPv4 and IETF RFC 1195, if an IS uses IS-IS protocol to support IPv6 routing, the IPv6 NLPID must be added to the NLPID TLV for issuing.

If there are two paths corresponding to a given prefix, adjustment must be considered in selecting the preferred path according to the up/down bit. The new order of preference is as follows:

- Level 1 up prefix;
- Level 2 up prefix;
- Level 2 down prefix;
- Level 1 down prefix.

If multiple paths have the same best preference, then selection takes place based on the metric. If the router supports equal-cost multipath routing, then the multiple paths should be treated as equal-cost multipath routing, otherwise the router can select any of the paths.

4) Next-Generation Routing Information Protocol (RIPng)

The routing information protocol (RIP) is one of the earliest IPv6-based dynamic IP routing protocols developed and used by the IETF, and the current versions are RIPv1 and RIPv2. In 1997 when the IETF developed the IPv6 standards, it made improvements to the RIP protocol in order to make it compatible with IPv6. It developed the RIPng (next generation) standards, the next-generation routing information protocol is based on IPv6 and defined in the IETF RFC 2080. RIPng is extremely widely used, and is one of the de facto standards for intra-AS routing protocols.

The goal of RIPng is not to create a brand-new protocol. Instead, necessary changes were made to RIP so that it can adapt to the routing requirements of IPv6. Therefore, the basic working principle of RIPng is the same as that of RIP. This is mainly reflected in the following aspects:

- The RIPng routing mechanism is based on the distance vector protocol;
- RIPng exchanges routing information through UDP packets. The UDP port used is 521 (RIP uses 520);
- RIPng sends an update packet every 30 seconds. If no update is received from a neighbor after 180 seconds, then the neighbor is deemed to be unreachable. If no update is received after a further 120 seconds, the neighbor is deleted from the route;
- In order to avoid loops, RIPng also uses split horizon and poison reverse techniques;
- Compared with RIPv1 and RIPv2, the major changes in RIPng are to the address and packet formats;

- Address space. RIPv1 and RIPv2 are based on IPv4. As such, the address space is 32 bits. RIPng is based on IPv6, thus all addresses used are 128 bits;
- Subnet mask and prefix length. RIPv1 is designed for subnet-less networks, so there is no concept of subnet masks. This means that RIPv1 cannot be used to propagate variable-length subnet addresses or unclassified addresses in CIDR. RIPv2 adds support for subnet routing, so subnet masks are used to distinguish between network and subnet routes. As an IPv6 address prefix has a clear meaning, the concept of subnet masks no longer exists in RIPng. Instead, the prefix length is used. For the same reason, as IPv6 addresses are used, there is no need to distinguish between network, subnet, and host routes in RIPng;
- Usage scope of the protocol. The usage scopes of RIPv1 and RIPv2 are not restricted to the TCP/IP protocol family. They can also fit the specifications of other network protocol families. Therefore, the routing table entry of the packet contains the network protocol family field, but the actual implementation procedure is rarely carried out in non-IP networks. Thus, support for this feature has been removed in RIPng;
- Representation of the next hop. In RIPv1, there is no information on the next hop. The receiving router regards the packet's source IP address as the next hop of the network route to the destination. In RIPv2, information on the next hop is explicitly given to facilitate optimal route selection, and to prevent loops and slow convergence. Unlike RIPv2, in order to prevent the RTE from being too long and to improve the efficiency of routing information transmission, the "next-hop" field in RIPng exists as a separate RTE;
- Packet length. In both RIPv1 and RIPv2, there are restrictions on the packet length, with each packet only allowed to carry a maximum of 25 RTEs. In RIPng, there is no restriction on either the packet length or the number of RTEs that each packet can carry. Instead, the length of the packet is determined by the MTU of the medium. This treatment of packet length in RIPng improves the network efficiency of routing information transmission;
- In RIPng, the address "FF02::9" is used for multicast updates;
- As RIPng runs on the IPv6 protocol, it relies on the IP authentication extension header and IP encapsulating security payload extension header defined in the IPv6 protocol to ensure protocol integrity and encryption authentication in the routing information exchange process. It no longer provides an identity authentication mechanism.

Please refer to IETF RFC 2080 for the specific implementation of the RIPng protocol.

3.5.2 External Gateway Protocol

1) Overview

The external gateway protocol is used between autonomous systems for the exchange of reachability information among a group of networks in a particular autonomous system and neighboring autonomous systems.

Currently, IPv4 networks mainly use the BGP4 inter-domain routing protocol. The BGP protocol is used to exchange network reachability information among BGP operators. The network information contains the complete AS list through which the traffic must pass to reach a certain network. The BGP protocol was originally designed only to broadcast inter-domain IPv4 routing information. The following three attributes specified in the protocol are explicitly related to the IPv4 protocol:

- NEXT_HOP (using an IPv4 address);
- AGGREGATOR (to store an IPv4 address);
- NLRI (using the IPv4 address prefix).

To enable BGP4 to support multiple network-layer protocols such as IPv6 and IPX, the IETF has formulated the IETF RFC 4760, which defines an extension mechanism for BGP4. This mechanism allows the BGP4 protocol to carry routing information of multiple network-layer protocols such as IPv6 and IPX.

In the abovementioned mechanism, it is first assumed that any BGP speaker (including speakers that support the multi-protocol extension function) has an IPv4 address. Therefore, for the BGP4 protocol to support multiple network-layer protocols for routing, the following two functions must be added to the BGP4 protocol:

- The newly-added information in BGP4 must associate a specific network-layer protocol with the next hop information i.e. the next hop address is represented using the specific network-layer protocol address;
- The ability to associate a particular network-layer protocol with NLRI. This requires the use of address families to distinguish between different network-layer protocols.

The NEXT_HOP attribute contained in the message must provide address information about the next hop only if the BGP4 protocol needs to issue a routing message for reachable destinations. When the BGP4 protocol issues routing messages for the removal of the unreachable destinations for certain routes from the server, the NEXT_HOP attribute does not necessarily provide address information for the next hop. Therefore, the reachable destinations and next hop address information in the routing messages under the BGP4 protocol should be combined and issued together. Additionally, the issue of routing messages for reachable destinations should be separate from that for unreachable destinations.

The multi-protocol extension specified in the IETF RFC 4760 is backward compatible. For example, a router supporting the multi-protocol extension can be compatible with a router that does not.

In addition, the IETF has also formulated the IETF RFC 2545 based on the abovementioned multi-protocol extension mechanism. This stipulates the support of the multi-protocol extension to the IPv6 protocol, thereby realizing the application of the BGP protocol in IPv6 networks.

The BGP protocol in IPv6 is usually referred to as the BGP+ protocol to indicate the difference.

2) BGP4 Multi-Protocol Extension

As shown in Table 3-2, the BGP4 multi-protocol extension primarily extends two new path attributes: MP_REACH_NLRI (multi-protocol reachable NLRI) and MP_UNREACH_NLRI (multi-protocol unreachable NLRI). Through these two attributes, the BGP4 routing protocol can issue routing information about various network-layer protocols such as IPv6 and IPX.

Table 3-2: List of newly-added attributes in the BGP4 multi-protocol extension

Attribute	*Type Code*	*Purpose*
MP_REACH_NLRI	14	To carry the set of reachable destinations. Used for forwarding these destinations.
MP_UNREACH_NLRI	15	To carry the set of unreachable destinations.

Both of these attributes are optional and non-transitive. When a BGP speaker that does not support the multi-protocol capabilities receives a BGP message containing these two attributes, the information in the attributes should be ignored and not passed on to other BGP peers. The multi-protocol extension is backward compatible, i.e. routers that support multi-protocol extensions can interoperate with those that do not.

(1) MP_REACH_NLRI attribute
The MP_REACH_NLRI attribute can be used for the following purposes.

- Issue a valid route to BGP peers;

- Allow a router to issue its network-layer address. The network-layer address is located in the "network-layer reachability information" field of the MP_NLRI attribute. The address is used as the next hop address of the destination.

The encoded format of the MP_REACH_NLRI attribute is shown in Figure 3-27.

Address Family Identifier (AFI) (2 bytes)
Subsequent Address Family Identifier (SAFI) (1 byte)
Length of Next Hop Network Address (1 byte)
Network Address of Next Hop (variable length)
Reserved (1 byte)
Network Layer Reachability Information (NLRI) (variable length)

Figure 3-27: Encoded format of the MP_REACH_NLRI attribute

The description of each field is as follows:

- **Address family identifier (AFI)**: This field is two bytes in length. It is combined with the SAFI field to identify the set of network-layer protocols to which the address in the next hop field belongs, the encoding of the next hop address, and the semantics of the NLRI field. If the next hop can come from multiple network-layer protocols, the encoding of the next hop must contain a way to determine its network-layer protocol.
- **Subsequent address family identifier (SAFI)**: This field is one byte in length. It is combined with the AFI field to identify the set of network-layer protocols to which the address in the next hop field belongs, the encoding of the next hop address, and the semantics of the NLRI field. If the next hop can come from multiple network-layer protocols, the encoding of the next hop must contain a way to determine its network-layer protocol.
- **Length of next hop network address**: The length of this field is one byte, and its value represents the length of the "Network Address of Next Hop" field. The length is measured in bytes.
- **Network address of next hop**: This field is of variable length and contains the network address of the next router on the path to the destination. The network-layer protocol and the "network address of next hop" in the attributes are

represented as <AFI, SAFI>.

- **Reserved**: This field is 1 byte in length and must be set to 0. The receiver should ignore this field.

(2) Network Layer Reachability Information (NLRI)

This field is variable in length and is used to list the NLRI for the valid routes issued by this attribute. The semantics of the NLRI are identified by the <AFI, SAFI> combination carried in the attribute.

The next hop address information carried in the MP_REACH_NLRI path attribute specifies the network-layer address used by the router. The router is the next hop of the destination(s) listed in the MP_NLRI attribute contained in the UPDATE message.

The rules for the address information of the next hop are the same as those for the information carried by the NEXT_HOP attribute under the BGP4 protocol.

If an UPDATE message contains the MP_REACH_NLRI attribute, it must also carry both the ORIGIN and the AS_PATH attributes regardless of whether it is in EBGP or IBGP message flow. Additionally, the UPDATE message must also carry the LOCAL_PREF attribute in an IBGP message flow.

If no other NLRI information is carried in the UPDATE message except that carried in the MP_REACH_NLRI attribute, then the UPDATE message should not carry the NEXT_HOP attribute of the next hop address. If a BGP speaker receives this type of messages with the NEXT_HOP attribute, it should ignore the NEXT_ HOP attribute in the message.

An UPDATE message should not contain more than one identical address prefix (i.e. the same <AFI, SAFI> combination) in the following fields: the WTHEDRAWN ROUTES field, the "network reachability information" field, the MP_REACH_NLRI field, and the MP_UNREACH_NLRI field.

(3) MP_UNREACH_NLRI attribute

This attribute can be used to withdraw unreachable routes from the router. The encoding format is shown in Figure 3-28.

Address Family Identifier (AFI) (2 bytes)
Subsequent Address Family Identifier (SAFI) (1 byte)
Withdrawn Routes (variable length)

Figure 3-28: Encoding format of the MP_UNREACH_NLRI attribute

The description of each field is as follows.

- **Address family identifier (AFI)**: This field is two bytes in length. It is combined with the SAFI field to identify the set of network-layer protocols to which the address in the next hop field belongs, the encoding of the next hop address, and the semantics of the NLRI. If the next hop can come from multiple network-layer protocols, the encoding of the next hop must contain a way to determine its network-layer protocol.
- **Subsequent address family identifier (SAFI)**: This field is one byte in length. It is combined with the AFI field to identify the set of network-layer protocols to which the address in the next hop field belongs, the encoding of the next hop address, and the semantics of the NLRI. If the next hop can come from multiple network-layer protocols, the encoding of the next hop must contain a way to determine its network-layer protocol.
- **Withdrawn Routes**: This field is variable in length and is used to carry the NLRI listed in the routes withdrawn from service. The semantics of the NLRI are identified by the <AFI, SAFI> combination carried in the attribute. If the UPDATE message contains the MP_UNREACH_NLRI attribute, then it does not need to carry any other path attribute.

(4) NLRI Encoding:

The network-layer reachability information (NLRI) field code comprises one or more two-dimensional arrays. The format of the two-dimensional array is {length, prefix}, and the encoding of the field is as shown in Figure 3-29.

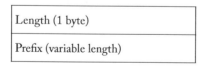

Figure 3-29: Encoding of the NLRI field

The description of each field is as follows:

- Length: This field is used to indicate the length of the "address prefix" field, and is measured in bits. When its value equals to zero, it means that the address prefix will match all addresses with the same address prefix as indicated by the "address family identifier" field.

- Prefix: The prefix field contains an address prefix. The address prefix is followed by enough padding bits. These padding bits make the length of the field an integral multiple of bytes. It should be noted that the value of the padding bits is irrelevant.

(5) *Subsequent address family identifier (SAFI)*

The values and meanings defined by the Subsequent Address Family Identifier (SAFI) field in both the MP_REACH_NLRI and the MP_UNREACH_NLRI attributes are as follows:

- Network-layer reachability information used for unicast forwarding;
- Network-layer reachability information used for multicast forwarding.

(6) *Error processing*

If a BGP speaker receives an UPDATE message containing either the MP_REACH_NLRI attribute or MP_UNREACH_NLRI attribute from a neighbor, and the receiver ascertains that the message contains incorrect attributes, then it must withdraw all BGP routes received from the same neighbor with the same AFI/SAFI as those carried in the incorrect MP_REACH_NLRI or MP_UNREACH_NLRI attributes. During the BGP session period in which the erroneous UPDATE message is received, the BGP speaker should ignore all routes with the same AFI/SAFI subsequently received during the session.

In addition, BGP speakers can terminate BGP sessions in which incorrect UPDATE messages are received. A NOTIFICATION message should be used to terminate the session. The message error code is "Update Message Error", and the error sub-code is "Optional Attribute Error".

(7) *Advertisement of BGP capabilities*

A BGP speaker using multi-protocol extensions should use capability advertisements to determine if it can carry out multi-protocol extended BGP capability interaction with a particular peer.

In the BGP "optional capabilities parameter" field, the "capability code" field is set to 1 to indicate that the multi-protocol extension capability is supported. The

Figure 3-30: Encoding of the capability field

"capability length" field is set to 4, and the capability field is encoded as shown in Figure 3-30.

The meaning and purpose of each parameter in this field are as follows:

- AFI (address family identifier) – 16 bits in length. The encoding method is as described above;
- Res (reserved field) – 8 bits in length. The sender of the message should set the value of this field to 0. The receiver of the message should ignore this field;
- SAFI (subsequent address family identifier) – 8 bits in length. The encoding method is as described above.

BGP speakers that support multiple <AFI, SAFI> arrays should use them as the multiple capabilities of the optional capabilities parameter.

In order to exchange the routing information specified by <AFI, SAFI> between a pair of designated BGP speakers, each BGP speaker must advertise its ability to support the route or routes specified by <AFI, SAFI> to its peer through the capability advertisement mechanism.

3) BGP4 multi-protocol extension support for IPv6

IETF RFC 2545 specifies the mechanism for transmitting IPv6 routing information using the MP_REACH_NLRI and MP_UNREACH_NLRI BGP attributes defined by the BGP4 multi-protocol extension.

(1) Basic requirements

Similar to other distance vector routing protocols in general, the BGP4 protocol is usually independent of the specific address family used by the protocol. It supports the IPv6 protocol.

In terms of routing information, the main differences between IPv6 and IPv4 protocols are the introduction of scoped unicast addresses in IPV6, and the definition of special circumstances in which a specific address range must be used.

(2) IPv6 address range

At present, two types of unicast address ranges are defined in IPv6, i.e. global addresses and link-local addresses.

The IPv6 specifications state that only link-local addresses can be used to generate ICMP redirect messages like ND, and next hop addresses such as RIP in some routing protocols. These restrictions imply that an IPv6 router must have a link-local next-hop address for all directly-connected routes, i.e. those in which the given router has the same subnet prefix as the next-hop router. However, according to the rule of the "next hop" attribute laid down in the BGP4 protocol specification, the link-local address is not suitable for use as a "next hop" attribute in the BGP4.

For the abovementioned reasons, when BGP4 is used to transmit IPv6 reachability information, at some points it is necessary to include both a global address and a link-local address when announcing the "next hop" attribute.

(3) Constructing the "next hop" field

A BGP speaker informs its peers in its "next hop network address" field of the global IPv6 address of the next hop, which is potentially followed by the IPv6 link-local address of the next hop.

When there is only one global address, the length of the "next hop network address" field in the MP_REACH_NLRI attribute should be set to 16. If the "next hop" field contains a link-local address, the value of this length should be set to 32.

The link-local address is only included in the "next hop" field when the BGP speaker shares the same subnet as the entity identified by the global IPv6 address carried in the "next hop network address" field, and the peer to which this route is published. In all other cases, the BGP speaker shall only advertise the global IPv6 address of the next hop (with the value of the "length of next hop network address" field set to 16) in the "network address" field to its peers.

Therefore, when a BGP speaker advertises a route to an internal peer, the network address of the next hop can be modified by removing the link-local IPv6 address of the next hop.

(4) Transport

BGP4 message exchanges occur over TCP connections, and TCP connections can be established based on both IPv6 and IPv4. BGP4 is independent of the transport protocol used, but obtains implicit configuration information from the address used to establish the peering session through the transport protocol. This information (i.e. the network address of a peer) should be taken into account during the route dissemination procedure. Therefore, when using an IPv4-based TCP connection to transport IPv6 reachability information, additional explicit network address configuration information about the peer is required.

The abovementioned information is different from the BGP identifier used in the process of BGP4 tie-breaking. The BGP identifier is a 32-bit unsigned integer included in an OPEN message. It is exchanged between two peers when establishing a session. This BGP identifier is a value that applies to the entire system. It is determined at start-up and must be unique in the network. At a given time, regardless of which network-layer protocol a particular BGP4 instance is configured to transport, the said BGP identifier should be obtained from an IPv4 address.

The use of IPv6-based TCP transport protocols to deliver IPv6 reachability information has the advantage of providing explicit confirmation of IPv6 reachability between two peers.

3.6 IPv6 Network Transition Techniques

In research conducted by the IETF, a variety of network migration techniques and methods have been proposed. These can be divided into the following three strategies: dual-stack strategy, tunneling strategy, and translation strategy. In practical networking applications, these strategies are seldom used alone. Usually, multiple techniques are combined and synergized to form an integrated network.

3.6.1 Dual-stack strategy

In dual-stack strategy, both the IPv4 and IPv6 protocol stacks are used in the network element. This means that the network is able to receive, process, send, and receive both IPv4 and IPv6 packets. For hosts (i.e. terminals), a "dual-stack" means that it can perform IPv4 or IPv6 encapsulation of data generated by the service according to requirements; for routers, a "dual-stack" means maintaining both the IPv4 and IPv6 routing protocol stacks in a routing device to enable it to communicate with both IPv4 and IPv6 hosts. Both IPv6 and IPv4 routing protocols are independently supported, and the routing information of IPv4 and IPv6 are calculated separately according to the respective routing protocols. Different routing tables are maintained. IPv6 datagrams are forwarded according to the routing table obtained by the IPv6 routing protocol, and IPv4 datagrams are forwarded according to the IPv4 routing protocol. The dual-stack structure is shown in Figure 3-31.

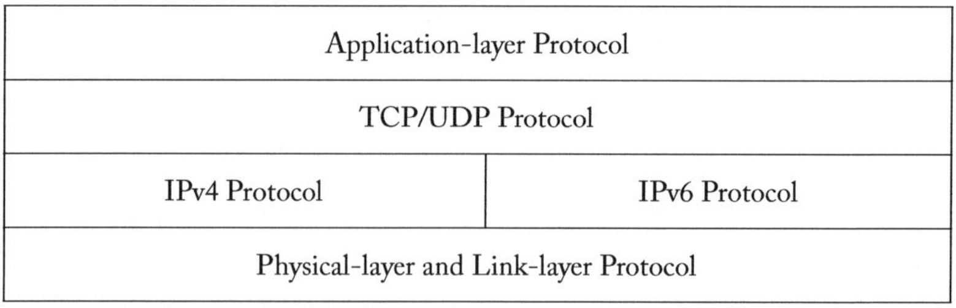

Application-layer Protocol	
TCP/UDP Protocol	
IPv4 Protocol	IPv6 Protocol
Physical-layer and Link-layer Protocol	

Figure 3-31: Dual-stack structure

1) Application method

In a network that supports IPv4/IPv6 dual stacks, a router is required to maintain the two sets of protocol specifications. In fact, a router hardware platform simulates two routers so that two logical networks are formed in a physical IP network. One is the IPv4 network, and the other is the IPv6 network. Both logical networks can cover the entire physical network at the same time. Since network elements that need to communicate can support both protocols, i.e. the network elements belong to two IP logical networks, at the same time, the communication protocol that is to be utilized can be selected among network elements through negotiation or according to the result of the domain name resolution. In this way, the network's support for the IPv6 protocol is realized.

An ideal case in which the dual-stack strategy is applied to networking is described above. However, in practical applications, greater flexibility can be accommodated, such as selectively applying the dual-stack strategy only to the regional core network and backbone network while the periphery network remains unchanged. Alternatively, the dual-stack strategy can be applied to the periphery network, while the regional core network and backbone network are still IPv4 networks.

For the former, user requirements for networking are relatively low. The user is free to select the communication protocol to be used, but the workload of upgrading and updating the routers of both the core and backbone networks is heavy. Another important problem that may occur in this case is the possibility of an IPv4 terminal being unable to communicate with an IPv6 terminal. Address/protocol conversion needs to be employed if communication is required.

For the latter, the operating system of the user terminal needs to be upgraded in order to support the dual-stack protocol, while the regional core network and backbone network can continue to use the original IPv4 protocol. This keeps the network structure and processing logic simple, thus ensuring that the network remains relatively stable and reliable. However, this brings about two other issues.

Firstly, the number of user terminals is large, and it is also the main network element type that consumes IP addresses. The use of the dual-stack protocol means that a user terminal may have both an IPv4 and an IPv6 address at the same time. As such, the number of IPv4 addresses saved is not as substantial as expected. Of course, to tackle this issue, the "temporary address" method (which is already being used in IPv4 networks) can be utilized to reduce the demand for public IP addresses to some extent, but this does not fundamentally solve the issue of insufficient IPv4 addresses. It also does not achieve the purpose of introducing IPv6 addresses to networks to gain a larger address space.

Secondly, this type of networking will require IPv6 terminals or subnets to communicate over IPv4. Currently, there are two ways of solving this problem: one is to use the tunneling technique; the other is to perform address/protocol translation twice. For tunneling, the IPv4 network acts as a transmission network for IPv6 data flow. There are many techniques to do this, such as manually-configured tunnels, 6to4 tunnels, 6over4 tunnels, and GRE tunnels. These will be discussed in detail in the following chapters. The other address/protocol translation technique is inefficient because it involves two-tier protocol translation from IPv6 to IPv4, and IPv4 to Ipv6. Thus, this method is not usually used in practical engineering.

2) Application features

In general, the advantages of the dual-stack strategy described above are that it is conceptually clear and easily understood, and network planning is relatively straightforward. At the same time, the logical IPv6 network is able to fully utilize all the advantages of the IPv6 protocol, such as security, routing constraints, and flow support.

However, the strategy also has the following disadvantages. The demand on network element devices is relatively high in that they have to support both IPv4 and IPv6 routing protocols. This requires the devices to maintain a large number of protocols and data. A possible solution would be a routing protocol that is compatible with the specifications of both IPv6 and IPv4 protocols, but no such protocol specification is available yet. In addition, network upgrading will involve all network element devices, thus requiring large investments and long construction periods.

3) Application range

As previously analyzed, the dual-stack strategy can be applied flexibly to all aspects of the network, such as hosts (terminals), network periphery (access and aggregation layers), and the core backbone layer. However, it is necessary to pay attention to the following issues when applying the dual-stack strategy at these network layers.

Firstly, relying solely on the dual-stack strategy (without the aid of tunneling and translation) to upgrade the network means that the entire network has to be unified. This requires a cutover of the whole network, failing which, certain parts would be unreachable by the router. However, a cutover of the entire network is unrealistic. As such, the dual-stack strategy is usually used on some part of the network in collaboration with other strategies, such as tunneling and translation, to jointly realize an integrated IPv4/IPv6 network.

Secondly, a network element adopting a dual-stack strategy has two protocol stacks but does not necessarily have both an IPv4 address and an IPv6 address concurrently.

Since the IPv6 address space is sufficiently large, each network element can have one or more IPv6 addresses, and is temporarily assigned an IPv4 address based on actual needs.

At present, the dual-stack strategy is used mostly in hosts (terminals) but rarely on its own. For example, in both the BIS and BIA techniques, the host (terminal) is required to support the dual-stack protocol.

For the network periphery, DSTM is the specific technique that can be applied, and it is based on the dual-stack strategy.

For the regional core layer and backbone network, the dual-stack strategy is not often used at present. Instead the original IPv4 network (sometimes combined with the MPLS technique) is used as the foundation, with the tunneling strategy adopted at the network periphery to provide a transmission channel for both IPv4 and IPv6 data.

4) DSTM—A typical technique

The DSTM (Dual Stack Transition Mechanism) is currently still in the draft stage at the IETF (as of 2003 it was still not an RFC). Nodes using the DSTM mechanism must be dual-stack nodes, and DSTM must be used with the tunneling technique.

(1) Applicable scenarios

The DSTM technique primarily solves the problem of how the dual-stack hosts (normally with only IPv6 addresses and not IPv4 addresses) in the IPv6 network communicate with the network elements (with only IPv4 addresses) in an external IPv4 network. The DSTM technique can only be applied to the internal network (i.e. an enterprise network or a residential network) and not to backbone networks or core networks.

(2) Basic principles

The purpose of DSTM is to provide the dual-stack nodes (with only IPv6 addresses) in IPv6-only networks (enterprise networks or residential networks) with a way to obtain an IPv4 address (a method to assign a temporary IPv4 address to dual-stack nodes), so that the nodes can make use of the IPv4-over-IPv6 tunneling mechanism to communicate with the IPv4 nodes or IPv4 applications in an external IPv4-only network, as shown in Figure 3-32.

(3) Functional components

The DSTM architecture includes a DSTM (address) server, a gateway or a tunnel endpoint, and several DSTM nodes.

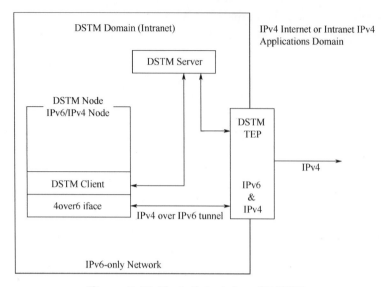

Figure 3-32: Basic Principles of DSTM

The DSTM server maintains an IPv4 address pool and is responsible for assigning IPv4 addresses (using multiple methods such as stateful DHCPv6 and stateless address auto-configuration) to the DSTM nodes according to their requests. The DSTM server only needs to guarantee that the IPv4 addresses assigned to the DSTM nodes are unique and valid for a certain period of time.

After obtaining the IPv4 address, the DSTM node encapsulates the IPv4 packet generated by the service in an IPv6 data packet before establishing a 4over6 tunnel to the tunnel endpoint (TEP). It then uses the IPv4-over-IPv6 technology to send the IPv4 packet to the gateway (TEP). Conversely, an IPv6 packet encapsulated with IPv4 data can be obtained at the TEP, and the de-capsulation of the IPv6 packet can be completed.

The gateway or TEP is the border router between an IPv6-only domain and the external IPv4 network. On one hand, it is responsible for establishing the 4over6 tunnel with the DSTM node; on the other hand, it de-capsulates the IPv6 data packet to realize the sending of an IPv4 data packet in the IPv4 network. In turn, data packets from the IPv4 network are accepted and transmitted to the appropriate DSTM nodes through the 4over6 tunnel.

(4) Notable issues

First, it is worth considering the basis on which the DSTM node decides to use IPv6 address for communication or to request the DSTM server to assign it an IPv4 address for communication. Generally, the IP address type of the destination address (or network element) is obtained through domain name resolution. How is the domain name resolution carried out in a network environment like this? It can be done in two ways, which usually correspond to two different application environments. In the first, there is a DNS server in the DSTM domain that can support the domain name resolution of the IPv4 address. At the same time, it can also resolve the domain name of IPv6 addresses, and can support two records: both A and AAAA. The DSTM node can use IPv6 to encapsulate the domain name resolution request and obtain the corresponding IPv4 address from the DNS server. Then, the DSTM node requests the IPv4 address from the DSTM server to communicate with the destination IPv4 network element. In the second application environment, there is no DNS server in the DSTM domain, or the DNS server only supports the domain name resolution of the IPv6 address (only AAAA records). In this case, translation and conversion of the DNS request message between DNS servers inside and outside the DSTM domain is required. The process is relatively complicated, and details can be found in the relevant DNS-ALG documents.

Secondly, since the IPv4 address assigned to a DSTM node by the DSTM server is temporary, the DSTM server will reclaim the IPv4 address after the usage time expires and there is no request from the DSTM node to continue using it. This means that the network element in the IPv4 network is unable to communicate with the DSTM node during this period, i.e. one-way communication from the DSTM node to the network element in the IPv4 network is possible but network elements in the IPv4 network are unable to initiate communication with DSTM nodes in the DSTM domain. The unidirectional communication is an important feature of the DSTM, which is actually a common problem for all communication methods using a temporary address assignment strategy. To solve this problem, a dynamic domain name resolution function has been proposed. A DSTM node has a fixed domain name while the IPv4 address corresponding to the domain name is temporary. In this way, the network element in the external IPv4 network can determine the temporary IPv4 address of the DSTM node through its domain name before engaging in bidirectional communication. Although this method solves the issue of bidirectional communication in a network application environment, it also increases the difficulty of managing and maintaining the domain name management system. Firstly, the DNS server needs to be able to request the DSTM server to assign a temporary IPv4 address to a DSTM node. This requires improvements to both the DNS server and the DSTM processing flow. Next, the domain name information in the domain name system is updated more frequently. This can cause instability of the system and may affect the consistency of the information. At the same time, the domain name system may be in a constant non-converged state, thus causing a void. The study of dynamic DNS is an important part of the current research into domain names.

In addition, performance considerations mean that multiple gateway TEPs may need to be set up for a DSTM domain. Information should be shared between these TEPs so that IPv4 traffic from one TEP can be supported, and the corresponding IPv4 traffic can be received from another TEP. In multi-TEP mode, the working mode of the network and the mode and content of the communication between TEPs are all worth studying. Research on this issue has not received enough attention at the IETF.

Last but not least, there may also be security loopholes in the DSTM technique. DSTM nodes can use TEFs as a relay to attack the network elements in an IPv4 network. In this regard, the IETF is of the view that the TEP can be placed inside an enterprise's or a residential area's network behind the firewall, so as to filter out some of the suspicious traffic within the network.

3.6.2 Tunneling

The tunneling strategy is a mechanism that is often used in IPv4/v6 integrated networking. The so-called "tunnel" is simply a technique that uses one protocol to transmit data belonging to another protocol. In an integrated networking environment, there are many bases for the classification of tunnel types.

According to the type of network element at the tunnel endpoint, tunnels can be divided into R-R (between routers), H-R (between a host and a router, which can be further divided into host to router and router to host), and H-H (between hosts).

Tunnels can be divided into manually-configured tunnels and automatically-configured tunnels according to their configuration method.

According to the types of tunnel and carrier protocols, tunnels can be classified into IPv4 over IPv6, IPv6 over IPv4, and IPv4/v6 over MPLS.

According to the direction of data flow, tunnels can be divided into unidirectional tunnels and bidirectional tunnels.

1) Applications

In an integrated networking environment, the tunneling mechanism can be applied in backbone networks and metropolitan area core networks. At the same time, it can also be applied to enterprise networks and customer premises networks.

During the process of building an integrated network (known as network transition), when adopting a periphery-to-backbone transition strategy, the backbone network or the metropolitan area core network continues to make use of the IPv4 protocol, while IPv6 networks or hosts exist at the periphery of the network. Communication between these IPv6 network elements can be carried out using tunneling mechanisms such as IPv6 over IPv4, or IPv4/v6 over MPLS.

In IPv6-only enterprise networks and customer premises networks, when an IPv4 host or an application needs to communicate with an IPv4 host or application in another network, it is necessary to use the tunneling mechanism to provide an IPv4 information transmission channel (IPv4 over IPv6) on the IPv6 network. Similarly, when an IPv6 host or application in an IPv4-only network puts up a communication request, the tunneling mechanism can also be used to provide an IPv6 information transmission channel (IPv6 over IPv4).

2) Basic principles

Tunnels include entrances and exits, known as tunnel endpoints. These tunnel endpoints are usually dual-stack nodes. At the tunnel entrance, the data of one protocol is encapsulated and sent in the form of another; at the exit, the received data is de-capsulated and processed accordingly. At the tunnel entrance, some information related to the tunnel is maintained, such as the MTU of the tunnel and other parameters. At the exit, the encapsulated data is usually filtered for security reasons to prevent malicious attacks from the outside.

The configuration of a tunnel can be carried out manually or automatically. Automatically-configured tunnels can be further divided into automatic IPv4-compatible tunnels, 6to4 tunnels, 6over4, ISATAP, MPLS tunnels, and GRE tunnels. Their implementation principles and technical details are not the same. As such, the scenarios in which each can be applied also differ.

3) Typical techniques

(1) Manually-configured tunnels

Manually-configured tunnels are mainly used when individual IPv6 hosts or networks need to communicate over IPv4 networks. This method is characterized by relatively straightforward implementation; the disadvantage is poor scalability. When there are many IPv6 hosts or networks that need to communicate, the workload for manual tunnel configuration and maintenance is heavy.

Manually-configured tunnels are suitable for use during the initial phase of integrated networks. At the same time, in the later phase of the integrated network, they can also exist as "default tunnels".

For a manually-configured tunnel, the IPv4 address of the endpoint is obtained from the configuration information (usually routing information) at the entrance. For every tunnel, the IPv4 address of the endpoint needs to be stored at the entrance. When an IPv6 data packet is transmitted through this tunnel, the endpoint IPv4 address of the configured tunnel will be the destination address of the encapsulated data packet.

(2) 6to4 tunnel

The 6to4 tunnel is a type of automatic tunnel. It is also a network transition mechanism to which the IETF attaches great importance, thanks to its broad application.

As for the 6to4 tunnel's main application environment, it is mainly used to allow isolated IPv6 subnets or IPv6 sites connected to IPv4-only networks to communicate with other similar sites prior to the establishment of IPv6-only connections.

The use of 6to4 tunnels to communicate is referred to as the 6to4 transition mechanism. In this way, IPv6 can achieve a high degree of independence with respect to a wide area network and can span many IPv4 subnets, allowing IPv6 "islands" in the "sea" of IPv4 to connect to each other. Multiple routing protocols (OSPF/BGP/RIP/ISIS) can be used in an IPv4 network. Routing between two 6to4 domains is made possible by the MP-BGP routing method. For the specific implementation of the 6to4 mechanism, only routers at the periphery need to be configured. As for the host, no other modification is needed beyond adding a default address option.

The 6to4 tunnel uses a special IPv6 address. IANA has permanently assigned an IPv6 format prefix of 0x0002 for the 6to4 transition scheme. The format of the IPv6 address prefix is 2002::/16. If a user site has at least one valid globally-unique 32-bit IPv4 address (v4ADDR), the user site will have the IPv6 address prefix 2002:V4ADDR::/48 without the need to submit an assignment request.

For routers at the periphery, IPv6 packets are encapsulated in IPv4 packets before being transmitted using tunnels. In the process of encapsulation at the start of the tunnel, the router at the periphery will extract the IPv4 address within the 6to4 address to be used as the address of the tunnel endpoint. The encapsulated IP packet is decapsulated when it reaches the destination 6to4 router. The IPv4 address of the site is included in the IPv6 address prefix, so the end of the IPv4 tunnel can be automatically extracted from the address prefix of the IPv6 domain.

In a transitional environment, there are two situations that need to be considered: one is that both parties in the communication are in a 6to4 domain, and use 6to4 addresses; the other is that one party is in a 6to4 domain and uses a 6to4 address while the other uses an IPv6 address in an IPv6-only domain. For the former, since the 6to4 routers of both parties can recognize and process the 6to4 encapsulation, the implementation is relatively simple. For the latter, the gateway of the IPv6-only network needs to be able to process both IPv6 data and 6to4 data, so that the gateway can act as a 6to4 repeater. The relay router participates in IPv6 unicast routing protocols on its IPv6-only interfaces, as well as on the pseudo 6to4 interfaces at the same time, even though both work in relatively independent routing domains. Repeaters can also participate in IPv4 unicast routing protocols at IPv4 interfaces that support 6to4. The networking modes corresponding to the two abovementioned cases are also slightly different.

The advantage of the 6to4 tunnel is that it is relatively simple to implement, and supports more devices. The disadvantage is that the 6to4 mechanism has a certain degree of coupling in the network placement, and a 6to4 repeater is needed for communication between a 6to4 domain and an IPv6-only one.

As a tunneling mechanism, the 6to4 tunnel also faces similar security issues. Furthermore, the existence of 6to4 repeaters further complicates the problem. Please refer to the IETF RFC for an analysis of the security of the 6to4 tunnel.

One of the main issues facing the 6to4 mechanism is the problem of route leaks. This refers to the leaking of routes in the IPv4 domain in 6to4 address formats to routes in the IPv6-only domain. In a networking environment that makes use of 6to4 repeaters, the relay routers must advertise a route to 2002::/16 to the IPv6-only external routing domain. How far this route advertisement to 2002::/16 can propagate in an IPv6-only routing system is of paramount importance. Choosing an incorrect strategy will lead to potential unreachability issues or poor transmission in the domain. In order to prevent the components of the IPv4 routing table from being introduced into the IPv6 routing table, 6to4 prefixes that are more accurate than 2002::/16 cannot be advertised in the IPv6-only routing domain. Therefore, a 6to4 site with an IPv6-only connection will not be allowed to advertise routes to 2002::/48 on this connection. All IPv6-only networks must filter out all 2002:: route advertisements whose prefix length is greater than /16.

(3) Automatic IPv4-compatible tunnel

The automatic IPv4-compatible tunnel is a type of automatic tunnel that is specified in the RFC of the IETF, but it is no longer recommended.

An automatic IPv4-compatible tunnel can be applied at the periphery of the network between dual-stack hosts or terminals (which can also be small IPv6 networks) in IPv4 networks that have IPv6 communication requirements.

IPv6/IPv4 nodes with automatic IPv4-compatible tunneling functions should be assigned an IPv4-compatible address. The IPv4-compatible address is composed of a 96-bit, all-zero prefix that is followed by a 32-bit IPv4 address. The node should only have an IPv4-compatible address configured when preparing to receive IPv6 data packets with an IPv4 address embedded in the destination address, which are encapsulated in IPv4 data packets. The destination address in the IPv4 encapsulated data packet header is equal to the lower 32 bits of the IPv4-compatible destination address in the IPv6 data packet header.

The destination address of the automatic IPv4-compatible tunnel is determined by the data packet passing through the tunnel. If the IPv6 destination address is IPv4-

compatible, the data packet can pass through the automatic tunnel. However, if the IPv6 destination address is a normal IPv6 address, the data packet cannot be sent through the automatic tunnel.

Routing table entries are able to guide data packets to the appropriate automatic tunnels. One way is a special static routing table entry for the prefix 0:0:0:0:0:0:0/96. This route uses an all-zero prefix as its 96-bit mask. Data packets match this prefix would be sent to pseudo interfaces that allow their automatic tunneling. Since all IPv4-compatible addresses match this prefix, all the data packets are sent to their destination through an automatic tunnel.

This automatic tunneling method is relatively simple to implement, unlike DSTM, which requires the dynamic assignment of IPv4 addresses for dual-stack nodes. However, its disadvantages include poor scalability (every dual-stack host needs to be assigned an IPv4-compatible address), and the need for a substantial number of IPv4 addresses (as the compatible address corresponds to an IPv4 address, so each dual-stack host that needs to communicate must be statically assigned an IPv4 address). Based on these reasons, the IETF does not recommend using the abovementioned method during network transition. IETF is also conservative on the usage of IPv4-compatible addresses.

As a type of tunneling technique, it also has certain security concerns. The tunnel endpoint is susceptible to attacks by unknown hosts. There are several ways to prevent this, such as filtering at both the entrance and exit of the tunnel, and IPSEC encapsulation.

(4) 6over4 tunnel

The 6over4 mechanism can only be applied to the periphery of the network, such as enterprise networks and access networks.

6over4 enables isolated IPv6 hosts that are not directly connected to IPv6 routers to form IPv6 interconnections by using IPv4 multicast domains as their virtual link-layer. In order to implement IPv6 routing, at least one IPv6 router that uses 6over4 needs to be connected to the 6over4 host in the same IPv4 multicast domain. Hosts that are interconnected using this mechanism do not require IPv4-compatible addresses or configured tunnels. By using this mechanism, IPv6 can be independent of the underlying links, and can span IPv4 subnets that support multicast.

At the initial stage of IPv6 transition, a site can configure the interface that connects an IPv6 peripheral router to an IPv4 domain to support IPv6 over IPv4, while the other interface that connects to an IPv6 domain is configured to IPv6. Any host that supports 6over4 in the IPv4 domain can freely communicate with these routers or IPv6 domains without manually configuring tunnels or hosts that are IPv4-address compatible.

The 6over4 mechanism requires that IPv4 networks support the multicast function. At present, most networks do not have the multicast function. Therefore, the mechanism is rarely used in practice. In addition, the multicast feature in IPv4 is a virtual-link layer for local transport. Therefore, the scope of its application is strictly limited. It is only applicable for communication between dual-stack hosts, and cannot connect an isolated node to a global IPv6 network.

(5) Tunnel brokers

Tunnel brokers are typically used for small, stand-alone IPv6 sites, particularly where IPv6 hosts that are independently distributed in the IPv4 Internet need to connect to an existing IPv6 network.

The tunnel broker (TB) provides a way of simplifying the configuration of the tunnel, thus reducing the heavy workload of tunnel configuration. The idea of a TB is to provide a dedicated server as a tunnel broker to automatically manage tunnel requests sent out by users. Through a TB, the tunnel connection between the user and the IPv6 network can be established easily, providing access to external IPv6 resources. In the early days of IPv6, this mechanism was highly beneficial in attracting more users to realize IPv6 connections conveniently and quickly. Additionally, it was also a very simple and convenient access method for IPv6 providers in the early stage.

(6) ISATAP

The Intra-Site Automatic Tunnel Addressing Protocol (ISATAP) mechanism is defined in the RFC of the IETF. It is usually applied at the periphery of the network, such as an enterprise network or access network. It can be used in conjunction with the 6to4 technique.

ISATAP enables dual-stack nodes in an IPv4 site to access IPv6 routers through automatic tunnels. It also allows dual-stack nodes that do not share the same physical link with the IPv6 routers to deliver data packets to the IPv6 next hop through IPv4 automatic tunnels.

The ISATAP transition mechanism uses an IPv6 address with an embedded IPv4 address. Regardless of whether the site uses a global or private IPv4 address, it can use the IPv6-in-IPv4 automatic tunneling technique within the site. The ISATAP address format can use either the site unicast IPv6 address prefix or the global unicast IPv6 address prefix, i.e. both intra-site and global IPv6 routing are supported.

(7) MPLS tunnels

MPLS tunnels are mainly applied to backbone networks and metropolitan area core networks. They connect IPv6 islands to each other, and are particularly suitable for operators that have already implemented BGP/MPLS VPN services. This transitional approach allows operators to provide IPv6 services externally without having to upgrade existing core networks to IPv6 networks.

An IPv6 site must connect to one or more dual-stack PEs running MP-BGP through CEs. These PEs exchange IPv6 routing reachability information among themselves through MP-BGP. IPv6 data packets are sent through the tunnels.

The advantages of this tunneling method are as follows:

Firstly, it is suitable for periphery-to-core network transition. Both the backbone and the metropolitan area core networks can maintain the original IPv4 protocol, while the MPLS technique is used at the periphery of the network for the transmission of IPv4 and IPv6 data packets.

Second, it has better scalability. When MLPS is implemented in the original network, each network at the periphery can independently decide upon the network transition time and networking mode. The networking mode of local networks is not affected by the MPLS tunnel mechanism.

The main disadvantage of this approach is that its implementation is based on the precondition that MPLS has already been deployed and implemented in the network, and so is not applicable to a network without MPLS deployment.

(8) Layer 2 tunnel

One possible way to connect scattered IPv6 networks is the usage of "layer 2" techniques such as ATM, PPP, and L2TP. This approach is mainly used to interconnect a small number of relatively important IPv6 networks (also known as islands).

The advantage of this method is that it is conceptually clear and easy to understand. The disadvantages are that it is difficult to implement and has poor scalability. When there are many IPv6 networks that need to be interconnected, this method is not recommended.

4) Main issues

The various tunneling techniques mentioned above have common problems that need to be analyzed in depth. There are important implications for the analysis of integrated networking.

(1) The MTU problem

In tunneling, a protocol (known as the bearer protocol) is used to transmit data encapsulated using another protocol (known as the payload protocol). There is an MTU value for the link formed by the bearer protocol (such as IPv4). When the size of the payload protocol data packet (such as IPv6) is greater than the MTU value of the bearer protocol after subtracting the length of the encapsulation format of the bearer protocol, segmenting of the payload protocol data packet will occur at the tunnel entrance, while reassembly will take place at the tunnel endpoint. This increases the complexity of implementing tunnel entrances and endpoints.

Specifically, when the bearer protocol is IPv4 and the payload protocol is IPv6, the MTU of IPv6 can be 65535-20 bytes (20 bytes is the length of the IPv4 header) only if the processing of data packets in IPv4 is considered. At the tunnel entrance, when the length of an IPv6 data packet exceeds this MTU value, an ICMP error message saying "Packet too Big" is sent to the source; when the length of the IPv6 data packet is less than this MTU value but greater than the MTU of the IPv4 transmission link, transmission is possible, but such a high MTU will cause some problems. First, it will result in the segmentation of IPv6 data packets. When the MTU of the IPv4 path minus 20 is less than 1280 bytes, data segmentation will still occur because in IPv6 specification, the MTU of any link layer must be greater than or equal to 1280 bytes. In the event that data packets go missing, the number of data packets retransmitted exceeds that which are missing, resulting in degraded performance. Therefore, excessive packet segmentation at the IPv4 layer should be avoided as far as possible. Second, data packets that are segmented in tunnels due to IPv4 need to be reassembled at the tunnel endpoints. If the endpoint is a router, the reassembly of data fragments takes up the router's memory. The IPv6 data packets are only forwarded after they are properly reassembled.

A possible way to solve the abovementioned problems is described here. First, the MTU value of the IPv4 path (tunnel) is obtained (please refer to the IPv4 path MTU discovery protocol for details). Next, this value is stored at the tunnel entrance (the IPv4 path MTU of the tunnel). After that, the IPv6 MTU discovery protocol will take this value into consideration to ensure that the length of the IPv6 data packet does not exceed the IPv4 path's MTU minus 20 bytes. This avoids the segmentation of IPv6 data packets, simplifies processing, and improves the efficiency of transmission.

(2) Security issues

All tunneling techniques face security issues, but these issues are even more complicated in automatic tunneling.

- Tunnel entrances should have certain filtering functions so that unauthorized data is prevented from entering and causing attacks to the networks beyond the tunnel endpoint.
- Tunnel endpoints should also have certain filtering functions so that unauthorized data is prevented from entering the local network.
- How to establish trust between tunnel entrances and endpoints is also an issue worth considering. There are many authentication methods for configured tunnels, and data can be sent under encryption, such as ESP and AH. However, the security issues become relatively more complex for automatic tunnels as it is impossible to authenticate prior to each data transmission. As such, other network elements in the network may disguise their identities by mimicking the network beyond the tunnel entrance, in order to attack the network beyond the tunnel endpoint.

(3) Routing issues

When it comes to tunneling, routing is one of the issues that needs to be analyzed in depth. In brief, tunnel routing problems include selection, leakage, and loopback.

- Amongst the various tunneling techniques, there are two types of routing selection methods. One occurs at the bearer protocol layer, and the other happens at the payload protocol layer. Choosing between the two can be done relatively independently, and can be carried out according to the size or structure of the network, or the organization of an autonomous system. For example, RIPng or OSPFv3 routing protocols can be used in IPv6 ASs, whereas BGP4+ routing protocol can be used between IPv6 ASs. The focus should be on how to use the bearer protocol to transmit the routing information of the payload protocol. At present, MP-BGP is used to solve this problem.
- When analyzing tunneling techniques, attention should also be given to the problem of route leakage. This term refers to the routing information of the bearer protocol when it enters the routing table of the payload protocol and vice versa. It causes problems such as expansion of the amount of routing information, management confusion, and decreased efficiency. The primary cause of route leakage is the use of certain special methods to represent the addresses in the tunneling mechanism, especially in the automatic tunneling technique. These include 6to4 addresses and IPv4-compatible addresses, which contain IPv4 address information. The transmission of routing information among these addresses can cause leakage into the IPv6-only routing information in IPv6-only networks, thus introducing IPv4 routing information into IPv6 routing

tables. A specific example of address leakage has been mentioned previously, in an analysis of the 6to4 technique.

One way to solve this problem is to strictly limit the propagation of IPv6 routes related to the special address types. For example, to prevent the components of IPv4 routing tables from being introduced into IPv6 routing tables, 6to4 prefixes that are more precise than 2002::/16 cannot be advertised in an IPv6-only routing domain, and all IPv6-only networks must filter out all 2002:: route advertisements with prefix lengths greater than /16.

– Attention should also be paid to the problem of address loopback when it comes to routing analysis of tunneling techniques. Normally, routing issues at both the bearer and payload protocol layer will have been taken into account at the routing protocol design phase, so routing loopback will not occur. However, in the automatic tunneling technique, the involvement of two types of IPv6 addresses (IPv6-only addresses and special IPv6 addresses) might lead to routing loopback issue. Routing loopback is a by-product of route leakage. When route leakage occurs, route information is mixed up, which results in routing loopback. Solving route leakage also tackles the routing loopback problem.

(4) QoS issues

With regard to QoS problems brought by tunneling techniques, an analysis can be carried out from the following aspects:

– In the tunneling technique, when IPv6 data is transmitted through an IPv4 tunnel, the IPv6 hop number is reduced by one after passing through the tunnel. However, in reality, the data has experienced multiple hops within the IPv4 domain. This increases the difficulty of QoS management at the IPv6 level.

– There are some QoS considerations in the IPv6 protocol. For example, a 4-bit priority area is defined in the IPv6 packet header, in which 16 levels of priority can be indicated. Next, a 24-bit flow label is defined in the IPv6 packet header. This allows the router to carry out the corresponding processing based on the flow identified, without having to check the address, port, or other information. However, the presence of the IPv4 tunnel restricts the effectiveness of the IPv6 QoS mechanism mentioned above, mainly because the QoS of the IPv4 tunnel cannot be guaranteed. A way to overcome the problem is to define the mapping relationships between the IPv6 priority field value and the IPv4 tunnel priority, so that high-priority IPv6 data can still be processed preferentially in the IPv4 tunnel, thus guaranteeing the performance of IPv6 data to some extent.

(5) Scalability issues

Various tunneling techniques require tunnel entrances and endpoints to be used in pairs. Generally, tunnel entrances and endpoints are situated in different networks. When a certain tunneling technique is chosen, both networks (at least the routers at the periphery) are required to support the technique. As such, there is a certain amount of coupling in selecting the networking technique for both networks. This coupling directly restricts the scalability of the technique. In order to support the interconnection of different networks, routers at the periphery of networks need to have more functions, but this would lead to higher costs and implementation difficulties.

5) Summary of techniques

At present, statically-configured tunnels and Layer 2 tunnels are used to connect individual IPv6 networks to larger IPv6 networks through IPv4 networks. However, due to difficulties in organization, management, and maintenance, large-scale usage is not feasible.

6to4 tunnels and MPLS tunnels can be used in backbone networks and metropolitan area core networks to interconnect different IPv6 networks (or host terminals).

The tunnel broker technique can be used at the periphery of the network to interconnect IPv6 networks and the core IPv6 network.

A combination of ISATAP and 6to4 is used to set up enterprise networks and private networks.

The 6over4 mechanism and the IPv4-compatible address automatic tunneling mechanism are not recommended for IPv4/v6 integrated networking.

3.6.3 Translation Strategies

The dual-stack strategy solves the problem of IPv6 coexisting with IPv4. The implementation principle places network elements (such as a host) with dual-stack protocols so that the most appropriate can be chosen for external communication according to the needs of the application. For a detailed analysis, please refer to the related content earlier in this book. However, during network transition, it is impossible to require all hosts or terminals to be upgraded to support the dual stack. There will certainly be communication needs between both IPv4-only and IPv6-only hosts in the network. The difference in protocol stacks means that translation between them is necessary.

There are two aspects to protocol translation. One is translation at the IPv4 and IPv6 protocol layers; the other is the translation between IPv4 and IPv6 applications.

There are two important issues to consider when analyzing the translation strategies: one is the location where the translation is carried out; the other is the impact of the translation strategy on the network structure of integrated networking.

For the former, since processing in the backbone network needs to be as simple as possible to improve processing efficiency, translation should not be carried out in backbone networks or core networks. This means that translation strategies should only be applied at the aggregation level or at the host terminal of the network. In the IETF, there are techniques for both translation at the aggregation level (NAT-PT, TRT) and host terminal level (BIS, BIA).

For the latter, since certain information or relevant processing parameters need to be preserved when translation processing is being carried out, the data originating from a certain translation location needs to return through it, otherwise the translation function for proper communication cannot be completed. At present, little research has been done on how to organize and coordinate multiple translation locations, but this is certainly an area worth investigating.

In addition, the translation strategy undermines the end-to-end communication between an IPv4 host and an IPv6 host. This affects the normal functioning of some services (such as security) and operations. There has been a certain amount of research on this issue, particularly on IPSEC NAT-traversal problems and NAT-traversal issues in various services. Some ways to resolve these problems have been proposed, but the solutions tend to be difficult to implement and have relatively poor scalability. Therefore, further research is needed in this direction, but it should be noted that the potential is not very high.

1) Typical techniques

Translation strategies can correspond to multiple implementation techniques. NAT-PT and TRT are mainly applied to the network aggregation layer; BIA and BIS target the host terminals.

(1) NAT-PT

The IETF proposed the Network Address Translator-Protocol Translator (NAT-PT), in which IPv4-only nodes can communicate with IPv6-only nodes. A basic assumption in using the NAT-PT is that there is no other way (such as a local IPv6 connection or various forms of tunneling) to allow the IPv6 nodes in the IPv6 domain to communicate with the IPv4 nodes in the IPv4 domain. This means that the NAT-PT transition mechanism applies only to the communication between IPv6-only nodes and IPv4-only nodes. It should not be used between an IPv6-only node and the IPv4 portion of a dual-stack node.

The NAT-PT is a direct conversion method, and enables communication without any alteration of the upper-layer protocol. The key equipment in this solution is the

NAT-PT gateway, which allows the mutual conversion of both the IPv4 and IPv6 protocol stacks, including protocols at the network layer and transport layer, and some protocols at the application layer.

Specifically, the NAT-PT includes the following three functional modules: the Stateless IP/ICMP Translator (SIIT) protocol translation function (IETF RFC 2765), the dynamic address translation capability of the NAT, and the corresponding Application Layer Gateway (ALG). The SIIT protocol translation function defines the mutual translation process between IPv4 packets and IPv6 packets, mainly by the mapping relationship of the fields, and the methods of determining the relevant parameters. However, there is no mention of how to obtain a temporary IPv4 address during the translation process. The dynamic address translation function of the NAT solves this problem. It maintains an IPv4 address pool and is in charge of assigning temporary IPv4 addresses. It also maintains information related to the translation process. Together, the collaboration between SITT and NAT complete the translation between IPv4 and IPv6 packets but not the upper-layer application information encapsulated in IPv4 or IPv6 packets. The Application Layer Gateway (ALG) was proposed to resolve this problem. Based on this application, the ALG can translate the upper-layer applications in the UDP or TCP. Generally, different applications require different ALGs such as DNS-ALG and ALGs for other operations. A combination of the three functions mentioned above will complete the translation of information from the application to the network layer. As such, the NAT-PT is a relatively comprehensive solution for interoperation between IPv6-only and IPv4-only nodes in a large number of commonly used applications.

The NAT-PT is a relatively sound solution for the issue of interoperability between IPv4 and IPv6. Its biggest advantage is that the various original protocols can communicate with the new protocols without modification. However, there are some restrictions to the application of this technique. Firstly, when it comes to the topological structure, it is required for the conversion of all data packets in the same session to be carried out at the same router. Therefore, the NAT-PT is more suitable for networks with only one router exit. Next, some protocol fields are unable to retain their original meaning completely during the conversion. In addition, the protocol conversion method lacks end-to-end security (i.e. the IPSEC traversal problem is still not thoroughly resolved).

(2) TRT

The Transport Relay Translator (TRT) is applicable in an environment in which IPv6-only networks communicate with IPv4-only networks. The TRT system is located between an IPv6-only host and an IPv4-only host. It carries out both {TCP, UDP}/

IPv6 and {TCP, UDP}/IPv4 mutual data translation. There are two types of transport relays: TCP relay and UDP relay.

The biggest difference between TRT and NAT-PT is that TRT acts as a relay and communicates with both parties at the TCP/UDP level as a proxy. For example, a TCP relay establishes a TCP connection with both TCP communication parties. All TCP communication on both sides is handled by the TCP relay. Meanwhile, the NAT-PT only acts as a translator rather than as a communication agent.

The main advantages of TRT are: no modification is required for both IPv6-only and IPv4-only hosts, and there is no need to consider either the PMTU or data packet segmentation.

The deficiencies of TRT are as follows: firstly, it only supports bi-directional transmission, and does not support the conversion of uni-directional multicast data packets; secondly, it is a stateful transport relay system located between two communicating entities. Even if multiple TRT systems are deployed in the same area, a transport layer session must still pass through the same TRT system; thirdly, the TRT system is unable to convert NAT-unfriendly protocols such as IPsec.

The following description assumes that all operations are initiated by an IPv6 host, and the destination is an IPv4 host. If a suitable address-mapping mechanism is used, TRT can also support IPv4 to IPv6 data services.

(3) BIS

The above-mentioned NAT-PT and TRT techniques are applied at the network aggregation layer, while BIS and BIA are applied at the host and terminal.

The BIS technique involves the addition of several modules (the translator, extension name resolver, and address mapper) to a dual-stack host to monitor the flow of data between the TCP/IP module and the network card driver, and to perform the corresponding mutual translation between IPv4 and IPv6 data packets.

When the current host is communicating with other IPv6 hosts, these IPv6 hosts are assigned some IPv4 addresses inside the current host. These addresses are used only within the current host. Moreover, the assignment process is carried out automatically through the DNS protocol. Therefore, the user does not need to consider whether the host at the other end is an IPv6. In other words, the host can use existing IPv4 applications to communicate with other IPv6 hosts, making it a dual-stack host capable of supporting both IPv4 and IPv6 applications, thereby expanding the scope of application of the dual-stack host.

Specifically, when the translator receives a data packet from an IPv4 application, it translates the IPv4 header to an IPv6 header before segmenting the converted data packet (since the IPv6 header is at least 20 bytes larger than the IPv4 header), and sending it to the IPv6 network. The reverse conversion is carried out when an IPv6 data packet is received from an IPv6 network, but the data packet does not need to be segmented. The extension name resolver is used to return a correct response to a request from an IPv4 application. The address mapper manages an IPv4 address pool, which can also contain private addresses. At the same time, the address mapper maintains a mapping table containing IPv4 and IPv6 address pairs. When the resolver and the translator need to assign an IPv4 address for an IPv6 address, the address mapper selects one IPv4 address from the address pool it manages, and dynamically records the mapping relationship between the addresses in the mapping table.

In addition, the BIS mechanism can coexist with other conversion mechanisms.

(4) BIA

In the BIA technique, an API translator is added between the Socket API module and the TCP/IP module of the dual-stack host, so that it can mutually translate between the IPv4 and IPv6 Socket API function. This mechanism simplifies the conversion between IPv4 and IPv6, but does not require the IP header to be translated.

For dual-stack hosts making use of BIA, the assumption is that there are both TCP/IPv4 and TCP/IPv6 protocol stacks on the local nodes.

When an IPv4 application on a dual-stack host communicates with other IPv6 hosts, the API translator detects the Socket API function in the IPv4 application and invokes the IPv6 Socket API function to communicate with the IPv6 host, and vice versa. In order to support communication between the IPv4 application and the target IPv6 host in the API translator, the IPv4 address pool is allocated by the domain name resolver.

2) Problems

A simple introduction to translation mechanisms is given above, focusing mainly on the implementation principles and application scenarios. In reality, the translation mechanism involves many other aspects, such as security, efficiency, and bi-directional communication. Some of these problems are common, and looking into them aids our understanding of translation mechanisms and prompts in-depth analysis of integrated networking techniques.

(1) Application Environments for Translation Strategies

The application environments for various translation techniques have been analyzed and can be summarized as follows:

- Translation strategies should not be carried out in core networks;
- Translation strategies can be carried out in the aggregation layer and at host terminals, and there is no contradiction;
- The application of a translation strategy is the last option and should be avoided where possible, because its impact on performance is high, and it limits the scalability of the network.

(2) MTU issues

One of the differences between IPv4 and IPv6 is that path MTU discovery is required in IPv6, but optional in IPv4. This means that IPv6 routers do not segment data packets. Only the sender will do so.

If an IPv4 node carries out path MTU discovery (which can be done by configuring the DF-bit in the packet header), then path MTU discovery will be performed from end to end, i.e. through the translator. In this case, both the IPv4 and IPv6 routers will send an "data packet too big" ICMP message back to the sender. When the ICMP error message sent by the IPv6 router passes through the translator, the translator translates these ICMP error messages into a form that the IPv4 sender can understand. In this case, only IPv4 data packets will arrive in fragments at the translator, and the corresponding translated data packet will have an IPv6 fragmentation header.

If the IPv4 sender does not perform path MTU discovery, the translator has to ensure that the data packet does not exceed the path MTU on the IPv6 side. Since IPv6 guarantees that data packets of 1280 bytes do not need to be segmented, an IPv4 data packet can be segmented so that it matches the 1280 bytes of an IPv6 data packet. This means that when the IPv4 sender does not perform path MTU discovery, the translator must always carry an IPv6 fragmentation header to indicate that the sender allows the fragmentation operation.

Chapter 4

The Development of IPv6 Technology

Chapter Highlights:
- *Global IPv6 development*
- *IPv6 development in China*
- *Solutions for IPv6 transition in China*

Overview

In recent years, the next-generation Internet industry has been developing at an accelerated pace around the world. Various developed countries and regions have rolled out plans and strategies, and are deploying national resources in order to seize a strategic foothold in the field. The construction of the next-generation Internet has transitioned from the experimental phase to the commercial deployment phase. Important standards, key equipment, software, and system-related R&D have reached maturity, so large-scale testing and deployment have begun.

4.1 Global IPv6 Development

4.1.1 Distribution of Address Resources

1) IPv4 address resources have been fully allocated, and China is facing a shortage.

On February 3rd, 2011, the global IP address assignment agency IANA (Internet Assigned Numbers Authority) announced that all the IPv4 address in its pool had been allocated. On April 15th, 2011, the IP address assignment agency in Asia, the APNIC (Asia-Pacific Network Information Centre), started allocating from its last /8 IPv4 address block. APNIC's policy is that each of its members will receive a maximum of one /22 IPv4 address block (or 1024 IPv4 addresses) each time henceforth. Other regional IP address assignment agencies including the RIPE NCC (Europe) and the ARIN (North America) are expected to finish allocating all available IPv4 address resources in 2012 and 2013 respectively.

As of the end of December 2011, the number of Internet users in China reached 513 million, but Internet penetration was only 38.3%. China owned a total of 330 million IPv4 addresses (excluding Hong Kong, Macau, and Taiwan)[1], placing it second among countries and regions. It accounted for 7.72% of all IPv4 addresses available worldwide, even though the IPv4 address per capita was only 0.24. According to a research report by the China Academy of Telecommunication Research at MIIT (Ministry of Industry and Information Technology), China needs 34.5 billion IP addresses in the next five years[2]. The severe shortage will cause a bottleneck in the development of the Internet of Things, the mobile Internet, cloud computing, and tri-network integration in China. Even though the extensive use of techniques such as NAT will delay the depletion of IPv4 addresses, the rapidly growing demand for applications cannot be met. Moreover, the use of such techniques will significantly increase network complexity and cause difficulties for management, leading to lower security for networks and information, and threatening the quality of service.

2) As the global allocation of IPv6 addresses speeds up, China ranks fifth in the world for number of applications

By the end of March 2012, a total of 196 countries around the world had applied for IPv6 addresses. The top five countries in terms of number of IPv6 addresses owned are Brazil, the United States, Japan, Germany, and China. The number of IPv6 addresses owned by China is 9410 (/32), which is approximately 1/7 of Brazil's and 1/2 of the United States'. It ranks fifth among countries and regions, and accounts for 5.66%[3] of the total number of IPv6 addresses allocated globally. Most of China's IPv6 addresses were obtained after July 2011. China Mobile obtained a /20 IPv6 address block in August 2011, and China Telecom obtained four IPv6 address blocks of sizes /21, /22, /23, and /24 respectively in December 2011. China Unicom obtained a /22 IPv6 address block in July 2011.

4.1.2 IPv6 Support Capabilities

Since the end of the 20th century, a large number of experimental IPv6 networks have been built around the world. The more sizeable ones include Internet2 in the United States, GÉANT2 in Europe, APAN in the Asia-Pacific, TEIN2 across Asia and Europe, and CNGI in China. These trial networks have achieved high-speed interconnections, and form a large-scale international next-generation Internet IPv6 trial network. In terms of commercialization, more than 10 operators in Japan have implemented IPv6 operations. Among them, NTT (Nippon Telegraph & Telephone) has established a global IPv6 backbone network and is also implementing IPv6-based IPTV services in Japan. Comcast, the largest cable TV operator in the United States, began offering IPv6 services to users from the second quarter of 2010. In Europe, Telefónica and France Telecom have also launched IPv6-based services.

As of the end of March 2012, the quantity of IPv6 addresses allocated globally was 3.13% of all IPv6 addresses, of which the quantity advertised was 19.37%. The number of globally active IPv6 BGP routes is 8641[4].

Figure 4-1 shows the growth of active IPv6 BGP routes.

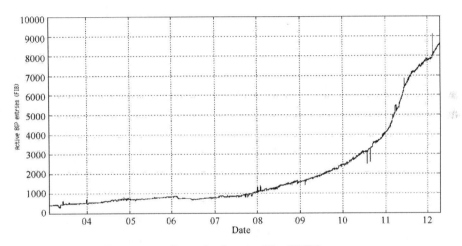

Figure 4-1: Growth of active IPv6 BGP routes

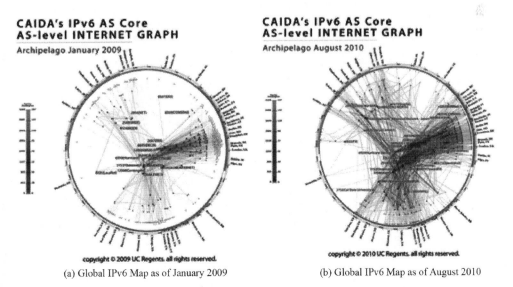

(a) Global IPv6 Map as of January 2009   (b) Global IPv6 Map as of August 2010

Figure 4-2: Global IPv6 Map

Figure 4-2 shows the topology of an autonomous IPv6 system (AS) published recently by the Centre for Applied Internet Data Analysis (CAIDA) at the University of California, San Diego (UCSD). As of August 2010, there were 21,852 IPv6 connections and 715 AS[5].

As of December 2011, nine of the 13 Domain Name System (DNS) root servers in the world had added valid AAAA records (records that point to IPv6 addresses), namely the A, D, F, H, I, J, K, L, and M root servers; 267 out of the 315 (or 84.76%) top-level domain name servers in the world support IPv6.

As of January 2012, more than 480 products worldwide had passed the IPv6-Ready Phase 1 certification, and over 670 products had passed the Phase 2 certification[6]. On the whole, there is now a wealth of IPv6 product types, including IPv4 product types, and they have been used at a considerable scale. A total of 131 Internet Service Providers (ISP) have passed IPv6-Enabled certification. Malaysia has the highest number of such ISPs – 13 in total – while four ISPs in China have been certified. A total of 1414 websites are IPv6 Enabled. China has the largest number of such websites – 306 in total – accounting for 21.64% of the total.

According to foreign media reports, a survey targeting IPv6 websites and conducted by the Measurement Factory (an Internet performance monitoring company) showed that the number of .com, .net, and .org websites supporting IPv6 technology has increased by 1900% over the last 12 months, accounting for 25.4% of the total – much higher than 1.27% of the total last year. This significant increase is mainly due to support for IPv6 by Go Daddy, the US domain name registrar.

The IETF is the main body for the international standardization of IPv6, and has formulated more than 200 IPv6-related standards. A system of core standards has been formed. At present, the bulk of its work focuses on transition techniques and refinement of existing standards. In addition to the IETF, other international organizations such as the ITU-T, 3GPP, and IEEE are also involved in the formulation of IPv6-related standards. The ITU-T focuses on the scenarios and requirements of applying IPv6 to NGN. The 3GPP focuses on IPv6 applications and 3G/LTE bearer and service applications. The IEEE focuses on the standardization of IPv6 applications in areas such as wireless LANs, smart grids, and energy conservation.

In addition, according to the 2011 Global IPv6 Deployment Monitoring Survey, more than 70% of the approximately 1600 international respondents (more than half of whom are Internet service providers) plan to deploy IPv6 for their networks by the end of 2012. The survey shows that positive progress has been made in the recognition of, planning for, and actual deployment of IPv6.

4.1.3 Attitudes of various countries to the development of IPv6

In the past two years, the development of the next-generation Internet industry has accelerated around the world. Some developed countries and regions have come up with national plans and strategies.

1) United States

The United States has adopted a military- and government-led IPv6 development strategy. The deployment of IPv6 is carried out concurrently in the areas of business, government, and the military. The Department of Defense, the Office of Management and Budget (OMB), and the Department of Commerce are responsible for each of the three areas.

In June 2003, the US Department of Defense issued a memorandum on the transition to IPv6, and formulated relevant policies for a Global Information Grid Program. According to the IPv6.com website, the US military has postponed the full deployment of IPv6 to 2012.

The OMB is responsible for the migration of government networks to IPv6. In September 2010, the US government issued an IPv6 action plan and published a detailed timetable for its IPv6 task force. All US government agencies are required to upgrade their public websites and services to support IPv6 (including Web, E-mail, DNS, and ISP services) by the end of 2012. By the end of 2014, the upgrading of application software related to the public Internet and enterprise networks should be completed.

The US Department of Commerce is responsible for the promotion of IPv6 in both commercial and private networks, with the NTIA (its subsidiary) responsible for implementation. In September 2010, the NTIA gathered companies such as Google, Verizon, Comcast, VeriSign, and Internet Society (ISOC) to discuss issues related to the promotion of IPv6. In addition, NTIA actively promotes the US broadband strategy, which supports the construction of demonstration networks and guarantees infrastructure for the development of IPv6 services.

September 30th, 2012 was the deadline for US federal agencies to support IPv6 on their public websites and online services. More than 10,000 websites were affected by this executive order, which has promoted the deployment of IPv6 in the United States.

2) The European Union

In May 2008, the European Parliament, the European Economic and Social Committee, and the European Committee of the Regions jointly issued the Action Plan for the Deployment of IPv6 in Europe, and requested that timely, efficient, and coordinated actions be taken within Europe to achieve the deployment goals. According to the EU's IPv6 roadmap, 25% of enterprises, government agencies, and home users will migrate to IPv6 by the end of 2010. However, this goal has not been achieved, and the utilization rate is about 8% in Europe

At present, the EU's policy measures to promote the development of IPv6 mainly include the establishment of demonstration projects through the development of large-scale testbeds to promote the tackling of interconnection and interoperability problems between IPv4 and IPv6; the promotion of government procurement to drive the development of IPv6, so that the government takes the lead in using IPv6; international cooperation, with the EU working with countries and regions such as the United States, Japan, and China on IPv6 projects, and constantly seeking new opportunities for cooperation.

3) Japan

In order to achieve a smooth transition to IPv6, the Japanese government has established a system for collaboration between the government and the people for co-promotion. IPv6 is an integral component in the "E-Japan" strategy. In August 2007, the Ministry of Internal Affairs and Communications set up the Internet IPv6-Transition Investigation and Research Committee to come up with Japan's IPv6 transition plan. In October 2009, the IPv4 Address Scarcity Work Group of Japan was jointly established by the Ministry of Internal Affairs and Communications, the JPNIC, and the Association of Telecommunication and Internet Operators, and the IPv6 Action Plan was issued. It was decided that IPv6 services would be launched fully in April 2011. Currently, there are 11 ISPs that provide commercial IPv6 services.

4) Korea

In September 2010, the Korean Communications Commission held a conference on the formulation of the Next-Generation Internet Protocol (IPv6) Promotion Plan, and announced that starting in June 2011, mobile communication services in Korea such as the Internet, IPTV, and 3G would make use of IPv6. In June 2011, the South Korean government announced that it had disabled IPv4 and fully deployed IPv6. However, there is no further confirmation that South Korea has successfully replaced IPv4 fully with IPv6.

On the whole, the pace of IPv4 to IPv6 transition varies greatly among countries and regions. According to the findings released by the IETF, the pace of transition is currently fastest in Japan. NTT, a Japanese telecommunication service provider, began providing commercial IPv6 services in June 2011. The US government plans to provide IPv6 services such as Web, E-mail, and DNS by the end of September

2012, but it will only provide IPv6 services such as basic networks and systems by the end of September 2014. Korea's ISPs are planning to provide IPv6 service in 2013. In Europe, France will begin with the switching of government department networks to IPv6 this year before providing IPv6-related services subsequently. On the whole, the transition to IPv6 in many countries will mostly take place in 2012 or later.

4.1.4 Collaboration with the Industry

Faced with the exhaustion of IPv4 address resources, the global Internet industry is working together to jointly promote the deployment of IPv6. On June 8th, 2011, more than 1000 companies and organizations such as ISOC, Google, Facebook, Yahoo, and Limelight Networks launched World IPv6 Day. For the first time ever, a 24-hour-long, large-scale IPv6 test was carried out globally, to promote the commercial application of IPv6 and identify potential technical issues.

On June 6th, 2012, another World IPv6 Day will be held, themed around "Global IPv6 Launch". According to the ISOC, ISPs including AT&T, Comcast, Free Telecom, Internode, KDDI, Time Warner Cable, and XS4ALL will all begin using IPv6 services on that day. In addition, websites such as Facebook, Google, Yahoo, and Microsoft's Bing will permanently enable IPv6 support. This means that IPv6 deployment has been officially launched around the world.

4.2 IPv6 Development in China

China's research into the next-generation Internet began relatively early. Following the implementation of a series of technological innovation programs, application demonstrations, and commercial trials at the national level, China has made some remarkable achievements. In terms of network construction, it has built the world's largest IPv6 network; in terms of R&D in equipment, it has acquired core manufacturing technologies for key IPv6 network equipment and fostered the relevant industries; in terms of business applications, it has carried out R&D in business systems, as well as demonstrations of applications of new services; in terms of technological innovation, it has made breakthroughs in network transition and security mechanisms. All of this provides China with a once-in-a-lifetime opportunity to make major leaps when it comes to the next-generation Internet.

4.2.1 Government roadmap and timeline for IPv6 development

The next-generation Internet has become an important element within China's emerging strategic industries. In the 12th Five-Year Plan for National Economic and

Social Development promulgated in March 2011, it was pointed out that there is a need to focus on the development of a new generation of basic information industries such as the next-generation Internet. This involves implementing relevant measures for the innovative development of strategic emerging industries, and promoting leapfrog development in key areas to realize transformation and upgrading and enhance core industrial competitiveness.

On December 23rd, 2011, the State Council's Premier presided over its executive meeting to discuss the deployment and accelerate the development of China's next-generation Internet industry. The meeting highlighted the need to seize this once-in-a-lifetime opportunity for technological change and industrial development. It encouraged innovation based on the existing mode, so as to develop a next-generation Internet that has adequate address resources, is advanced and energy-saving, is safe and credible, and has good scalability and a well-developed business mode. The meeting also clarified the roadmap and major goals for a small-scale commercial trial of IPv6 networks to begin by the end of 2013, to come up with a well-developed business model and a route for technological evolution. Between 2014 and 2015, deployment and commercial usage will take place on a large scale to implement the interoperability of mainstream IPv4 and IPv6 services.

In order to implement the recommendations arising from the State Council's executive meeting, eight ministries and commissions including the National Development and Reform Commission (NDRC) jointly came up with the Opinions on the Development and Construction of the Next-Generation Internet for the Period of the 12th Five-Year Plan (hereinafter referred to as "Opinions"). The development goals for China's next-generation Internet for this period are clearly stated in the Opinions. These include:

- Internet penetration in excess of 45%, tri-network integration;
- more than 25 million users with IPv6 broadband access;
- the interoperability of mainstream IPv4 and IPv6 services;
- and the acquisition of more than 10% of the total number of IPv6 addresses allocated globally.

Other goals include:

- high-end breakthroughs in the areas of theoretical research, software R&D, equipment manufacturing, and applications and services;
- significantly better support for the network from services, applications; and terminal equipment;

- to establish a system of domestic and international standards based on independently-developed technologies;
- to increase the number of international standard proposals by more than 75%;
- to establish a relatively complete security system for network and information safety;
- to significantly improve network and information security, regulatory capabilities, and management discourse power;
- to reduce the energy consumption per network information flow unit by more than 40%;
- to reduce the energy consumption per RMB$10,000 increment in value of the network equipment manufacturing industry by at least 15%;
- to allow independently-produced products in key areas to enjoy a domestic market share of 80% and above'
- to form a group of next-generation Internet research agencies and major companies with greater influence globally, which would create 3 million new job opportunities, enhance the pull effect on consumption, investment, and export, provide knock-on benefits, and play the leading role in the information industry, high-tech service industry, and social-economic development.

To achieve these goals, the NDRC announced two major projects – the Next-generation Internet Information Security Project, and another special project for technological R&D, industrialization, and large-scale commercial use of the next-generation Internet.

4.2.2 National projects to support and promote the development of IPv6

China attaches great importance to the development of the next-generation Internet, and relevant commissions and ministries have come up with a series projects related to it.

1) High-tech industry development projects

As part of the NDRC High-tech Industry Development Project, in 2003 the NDRC paired with ministries and commissions such as the Ministry of Education, the Ministry of Science and Technology, the Ministry of the Information Industry, the Chinese Academy of Sciences, the Chinese Academy of Engineering, and the National Natural Science Foundation of China to initiate the China Next-Generation Internet (CNGI) Demonstration Project. Six main backbone networks (covering 22 cities nationwide and connecting 59 core nodes), two domestic/international exchange centers (in Beijing and Shanghai), and 273 customer premises networks were built and deployed. In 2005 and 2006, a total of 103 CNGI technical testing and industrialization projects were established, of which 56 were for technical testing, application demonstrations, and standards research, and the remaining 47 were for system R&D and industrialization. Starting from the end of 2008, projects for the commercial testing of CNGI were implemented, including major schemes such as the Technical Upgrade and Demonstration for Education and Research into IPv6 Infrastructure, which is part of China's efforts to boost domestic demand, as well as 46 others for the commercial testing of services and industrialization. Domestically, tens of thousands of people from up to 100 colleges, up to 100 scientific research institutions, all national telecommunication operators, dozens of equipment manufacturers, and software developers participated in the abovementioned projects for the construction of the CNGI demonstration network. The aims are to provide a testbed for major scientific research and new services, to promote the formulation of standards and the industrialization of network equipment produced domestically, to obtain a large number of demonstration outcomes, to enhance China's innovation capabilities in the field of the next-generation Internet, and to train and cultivate a batch of next-generation Internet professionals.

2) Projects for building innovation capacity

In the NDRC's project for building innovation capacity, a national engineering laboratory focusing on next-generation Internet technologies such as core networks, access systems, interconnected equipment, and broadband service applications has been set up to provide innovative technologies, standards, and personnel for the development of the next-generation Internet. It is a platform for R&D and innovation in generic technologies.

3) Major science and technology projects

Topics of interest in research and industrialization such as system architecture evolution (SAE), mobile service control network, new IP bearer network architecture, development of new service applications, and R&D in key technologies in terminals and subscriber cards have been incorporated into the New-generation Wireless Broadband Mobile Communication Network project. This major project started off with top-down design and master planning, and includes research topics such as ubiquitous wireless network architecture and overall design, overall mobile Internet architecture, mobile application platform architecture, and mobile network and information security architecture.

4) Key science and technology projects

The Ministry of Science and Technology initiated a key technological engineering project known as the High-Performance Broadband Information Network (3TNet) during the 10th Five-Year Plan period. Transmission, routing, and switching equipment that were independently developed and able to support terabit networking were used to build an operational, high-performance broadband information demonstration network that supports large-scale concurrent streaming media services and interactive multimedia services in Shanghai (later extended to the Yangtze River Delta region). At present, the 3TNet is used to develop the architecture and key technologies for China's next-generation broadcasting (NGB) network, which is suitable for tri-network integration, the integration of wire and wireless networks, and integrated services.

During the 11th Five-Year Plan period, the Ministry of Science and Technology launched the New-Generation Highly-Trusted Network project, which includes the 863 program. Through projects such as the Future Packet-based Network (FPBN), a cross-regional testing network has been built for China's new network technologies that is flexible, reconfigurable, and supports tri-network integration. The Trusted Internet technological support program has also been launched, which focuses on research into a new generation of trusted Internet architectures and key technologies based on real IPv6 source addressing, to build a large-scale, trusted Internet testing network. The New Interactive Media Network and New Services Technological Project is a technological support program with broadcasting service at its core. It aims to build a demonstration model in order to form a new interactive media network that is bi-directional, multi-functional, and broadband-based.

The Ministry of Science and Technology's 973 Program supports relevant next-generation Internet research. Its primary focus in 2003 was on theoretical studies of architecture. Follow-up studies include "Basic Research into New-Generation Internet Architecture and Protocols" in 2009, and "Basic Research into Integrated Trusted Networks and Universal Service Systems" in 2007.

5) Basic research projects

During the 10th Five-Year Plan period, China's National Natural Science Foundation launched major research programs such as "Network and Information Security" and "Research into Scientific Activity in a Network-based Environment", to fund projects on next-generation Internet architecture, and basic theories and key technologies for the new-generation network application platform and network management.

6) Development fund for the electronic information industry

During the 11th Five-Year Plan period, the development fund for the electronic information industry was used to address key difficulties encountered in the development of the Internet, such as security, credibility, and controllability. The fund was also used to support research projects on the architecture and key technologies of future packet-switched networks, and R&D in equipment systems and network experiments.

4.2.3 The world's largest IPv6 demonstration network

Through the implementation of the CNGI project, a large-scale next-generation Internet demonstration network was established. It includes six backbone networks, two international switching centers, and 273 customer premises networks. The six backbone networks are the China Education and Research Network, and five others built by China Telecom, China Unicom, China Netcom/the Chinese Academy of Sciences, China Mobile, and China Railcom. CNGI international/domestic exchange centers were established in Beijing and Shanghai to interconnect the six backbone networks, and to connect to the next-generation Internet in the United States, Europe, and the Asia Pacific region. Nationwide, IPv6 customer premises networks have been built in 100 colleges, 100 research institutes, and 73 enterprises. CNGI-CERNET2, which was established by 25 colleges such as Tsinghua University, is the largest IPv6-only Internet in the world. It has achieved a number of major breakthroughs, and has reached world-leading levels.

On this basis, the CNGI demonstration network provides a testbed for major scientific research and new services. It provides a testing platform for the technological innovations and outcomes arising from China's research into next-generation Internet technology, its formulation of standards, and its development of products. It is also a testing environment for technological R&D in the National Natural Science Foundation of China's programs such as 973, 863, and Technological Support. It also promotes the implementation of major projects for the national scientific research programs managed by the Chinese Academy of Sciences and the Ministry of Education. In addition, various telecommunication operators have participated in the construction of the demonstration network, and have launched application trials and technical studies of new telecommunication services. In so doing, they have gained a great deal of valuable experience.

The High-performance Broadband Information Network 3TNet, built in the Yangtze River Delta region, has backbone networks capable of terabit transmission, switching, and routing capabilities. A year-long trial operation of the network involving 30,000 users in the region has been conducted. During the trial period, six special trials and large-scale simultaneous testing involving thousands of participants were also carried out. A follow-up demonstrative application involving a million users will be carried out in 15 cities or regions in the Yangtze River Delta.

Construction of the CNGI allowed enterprises and research institutes to obtain plenty of findings from demonstrative applications, some of which have contributed to the development of China's economy and society. An example is the use of the CNGI to establish a high-speed connection between China and Europe so that high-resolution, remote-sensing satellite images could be sent from the European Joint Research Centre to the Centre of Earth Observation of the Chinese Academy of Sciences in real time in the aftermath of the devastating Wenchuan earthquake in Sichuan Province in 2008. During the Beijing Olympics, the CNGI center in Beijing set up a mirror site of the official Olympics website, which was IPv6-based. This was an important demonstration of IPv6 application by China to the rest of the world. An IPv6 video surveillance and sensor system developed with the support of the CNGI project was successfully used at the 48 sports venues across the country, and the task of guaranteeing communication for the Olympic Games was executed smoothly. This had many positive repercussions internationally. In addition, China's environmental monitoring system based on IPv6 wireless sensor networks aims to develop a water pollution monitoring system for the Huaihe River basin, and provides advanced means for comprehensive watershed management and disaster warning. The Application Demonstration for Cooperation in Resources and Environmental Security Zones in the Lancang-Mekong Sub-region aims to develop and establish a testing platform for ecological monitoring in the region. The Hunan Agricultural Comprehensive Monitoring System based on the IPv6 Internet of Things and built by domestic telecommunication operators has received the first IPv6 Enabled ISP certification worldwide. Major operators in China have responded to the government's call. As well as participating in the construction of CNGI and trial engineering projects, they have also launched many next-generation Internet promotion efforts that have borne fruit. Some success has been obtained, and a series of application trials have been rolled out for agricultural informatization, smart homes, video information services, home entertainment, unified messaging, and secure VPN. For example, domestic operators in China provided IPv6 networks and related services at the Resource-saving and Environmentally-friendly Society in 2009, and at the 2010 Shanghai World Expo. IPv6 networks will be built and IPv6 business operations carried out at the 2011 Shenzhen World University Games.

4.2.4 Shaping a more complete system of standards for IPv6

Work on IPv6 standardization in China began in 2001. The China Communications Standards Association (CCSA) was placed in charge, and more than 40 standards have now been completed or are being developed. The outcome is that a relatively complete IPv6 standard system has been formed. The formulation of IPv6 standards in China has gone through three stages. The first was the localization of international standards, in which the formulation of IPv6 standards such as basic protocols and routing protocols was completed. The second was the formulation of a series of standards for technical equipment and testing, to meet the needs of IPv6 network construction in China. The third was the formulation of technologically innovative standards. Certain breakthroughs have also been made in terms of transitional technical standards.

Driven by the CNGI demonstration project, China has made great progress in international Internet standardization in recent years. This has allowed it to seize the initiative and have a greater say in the IETF. At present, four IETF task forces are helmed by Chinese experts. There is "Hokey" in the area of security, "MIF" in the area of multiple interfaces, and "PPSP" and "Decade" in the area of P2P. At the same time, experts from China have helped to bring about the establishment of task forces such as SAVI and Softwire, and have served as technical advisors in the SAVI and Softwire task forces. China has played an important role in location identification separation, mobility management, the Internet of Things, multiple interfaces, security, P2P, network transition, Chinese domain names, and Chinese email. At present, China's experts have led the formulation of more than 30 IETF standards. To a certain extent, this has changed China's long-term predicament of only being a passive receiver when it comes to core Internet technologies.

Overall, China is still a follower of international standards when it comes to IPv6. Its progress in IPv6 standards is basically aligned with international levels, but it has managed to innovate in terms of transition standards (such as Softwire and IVI technical standards), which have become international. At present, the IPv6 standards formulated by China are able to meet the construction needs of IPv6-only networks, but much more attention needs to be paid to IPv4/IPv6 interoperability standards and application standards, which are still work in progress.

4.2.5 Substantial progress made by the IPv6 industry

IPv6-related industries mainly involve network devices, service devices, mobile terminals, application software, operating systems, and chips. During the construction of the CNGI demonstration network, China used domestically-produced equipment as its mainstay. This propelled the development of its domestic industries and accelerated the industrialization of its next-generation Internet core equipment. At present, more than 50% of the equipment used in the CNGI demonstration network is domestically made. Routers, switches, broadband access equipment, Internet gateways, audio/video surveillance camera terminals, and wireless sensor network nodes have been mass produced and sent to market. China's R&D in equipment and its industrialization capabilities are also on par with advanced international levels. In the field of IPv6 testing, China has participated in the formulation and implementation of global test specifications. It provides leadership for IPv6-enabled testing and certification, and a Chinese expert is serving as the chairperson of the said task force. Under China's leadership, 60 ISPs and 494 websites worldwide have received global IPv6-enabled certification. In general, China's strengths in the IPv6 industry are in network equipment and service equipment. However, it is relatively weaker in terms of operating systems[7] and chips[8].

In terms of network equipment, the product types are as varied as IPv4 products (including routers, switches, and broadband access servers), and keep pace with foreign countries. Chinese equipment has also been deployed in the CNGI demonstration network. However, there is still some room for improvement of IPv4/IPv6 interoperability equipment (such as CGN) to meet the telecom-grade requirements for commercial-scale deployment. Service network equipment comprises mainly of periphery devices such as SIP servers, integrated access devices, enterprise gateways, and home gateways. Conventional core network devices such as soft switches, relay media gateways, signaling gateways, IMS systems, base station controllers, 3G packet switched domain subsystem, and intelligent network systems are relatively poorer in terms of IPv6 support capabilities. Although the various manufacturers have yet to roll out high-capacity IPv6 core network equipment, they have completed the relevant research for data network products. In addition, some core network devices already support IPv6 at the signaling level. Once there is a market for IPv6 core network equipment, the corresponding IPv6 products can be quickly launched.

As for information appliances, Chinese companies such as Haier, Konka, and Hisense have begun R&D into IPv6 terminal products, but the product types that support IPv6 are still relatively few. In the field of mobile smart terminals, as mobile phone chips and operating systems are mostly monopolized by foreign companies, its support to IPv6 is basically synchronized with that of the rest of the world. At present, there is still no commercially-available dual-stack phone that supports IPv4/IPv6.

Most operating systems are monopolized by foreign software vendors. At present, all PC operating systems (including Linux, UNIX, and Windows) support IPv6, but the client needs to be configured and enabled. As for mobile phone operating systems (including Android, ioS, RIM, Symbian, and Windows Mobile), even though the mainstream vendors all claim to be able to provide versions that support IPv6, most pre-installed versions do not support the IPv4/IPv6 dual-stack.

In terms of application software, development lags far behind that of operating systems. Currently, most software that is based on the Windows platform (such as QQ, Xunlei, and some game client software) does not support IPv6, but browsers do.

As for chips, the main breakthroughs by Chinese manufacturers are in TD chips (by Spreadtrum) and WCDMA chips (by HiSilicon and MediaTek), but the number of models supporting IPv6 is still relatively small[9].

4.2.6 The main problems facing the development of IPv6

At present, the following problems are hindering the development of IPv6 in China.

1) A lack of core drivers for full commercial deployment

The primary impetus for the deployment of IPv6 is the scarcity of IPv4 addresses, but various problems currently faced by the Internet remain, such as service quality and security. There is also no effective business model to generate economic benefits, yet a lot of capital investment is required. Therefore, there is insufficient financial motivation for many parties in the industry chain. The full commercial deployment of IPv6 cannot be achieved solely with action from a particular segment in the industry chain. As such, IPv6 deployment in China is unbalanced, i.e. strong in the middle (the network) but weak at both ends (the terminals and applications).

2) IPv6 support varies from segment to segment of the operators' networks, and preparation for the transition to the next-generation Internet is inadequate

The support of operators' networks for IPv6 can be divided into two parts. First is the network itself, including support for IPv6 by the access networks, metropolitan area networks, and backbone networks; the other is the back-end service support systems, including the billing system, network management system, DNS, and authentication systems.

Network:

- For backbone networks, router equipment provides decent support for IPv6. Most hardware does not pose a problem, and only the software needs to be upgraded;
- For metropolitan area networks, most BRAS (broadband remote access servers) and SR (service routers) only have minor hardware problems, but most of the software needs to be upgraded. In particular, a lot of BRAS software does not yet

support PPPoE for IPv6. There is no problem with the GGSN/PDSN hardware of the packet switched domain, but software upgrades are required;

- For access networks, transmission and switching systems can be treated as purely Layer 2, while Layer 3 functions can be handled by the BRAS or SR Gateway. Therefore, during the initial deployment phase of IPv6 services, whether or not IPv6 is supported has little impact on services. As home gateway routers generally do not support IPv6, they need to be upgraded and replaced.

From the perspective of back-end service support systems, the overall support for IPv6 is still relatively weak. Large-scale transformation of the entire system is required, including the billing system, network management system, authentication systems, and DNS.

The three domestic network operators each have a CNGI commercial-testing network. IPv6 service trials and application demonstrations have been carried out, but on the whole, the operators still do not have a comprehensive transition plan to solve the interoperability of IPv4 and IPv6.

3) Network transition products are still unable to meet the needs of telecom-grade commercial deployment

Commercial deployment of IPv6 networks must address two key issues: the coexistence of IPv4 and IPv6 networks, and the interoperability of IPv4 and IPv6 services. The relevant task forces in the IETF have come up with three technical solutions for transition. They are dual-stack, tunneling, and translation. Dual-stack and tunneling are used to tackle the issue of network coexistence, and both techniques are relatively well developed. Translation is used to tackle service interoperability between IPv4 and IPv6, and is less developed. At present, there is no mechanism that can completely solve the problem of coexistence and interoperability of IPv4 and IPv6 networks. During the network transition period, multiple mechanisms must be comprehensively applied according to the actual application scenario, which increases the complexity of network deployment.

At present, operators have a clear transition plan to overcome the coexistence issue of IPv4/IPv6 networks, i.e. dual-stack private networks (NAT44) and lightweight tunnels (DS_Lite). NAT444 makes use of the existing NAT technique to implement the two-level translation of IPv4[10] private network addresses by introducing a large-capacity, telecom-grade NAT device (CGN). At the same time, NAT444+IPv6 is used to form a dual-stack, private network architecture to realize the transition to IPv6. DS-Lite builds on the existing NAT technique to introduce IPv4-in-IPv6 tunnels to form a NAT+tunnel+IPv6 architecture to realize the transition. The main idea behind DS-lite is to use the IPv6 tunnel to avoid the complexity of traversing the NAT twice. Planning and management of private network addresses for users are not required. In terms of equipment, mainstream equipment manufacturers support both approaches, but the equipment is still some way off the telecom-grade requirements for commercial-scale deployment.

Translation is the only technique currently available to solve the issue of service interoperability between IPv4 and IPv6. Typical techniques include NAT64 and IVI. NAT64 is an IPv6-to-IPv4 conversion technique. Its main consideration is for IPv6 terminals to access IPv4 resources during the initial transition period, and not vice versa. The IVI technique uses stateless IPv4/IPv6 address mapping and a stateless protocol translation mechanism to implement two-way exchanges between IPv4 and IPv6, but does not address the issue of IPv4 address shortages[11]. In terms of equipment, neither techniques are yet mature, especially in terms of meeting the performance and reliability requirements of telecom-grade applications.

4) Challenges to data security; improvements to supervision technologies and products

IPv6 has a larger address space and simpler IP headers than IPv4, but the IP architecture of the two are fundamentally the same. As such, IPv6 essentially faces the same security risks as IPv4, as well as most of the security problems in IPv4 such as eavesdropping, application layer attacks, man-in-the-middle attacks, and flood attacks. In addition, because the IPv6 protocol stack is integrated into the IPsec protocol, a more convenient end-to-end encryption method is provided for terminal applications.

The expansion of address space and the use of IPsec encryption in IPv6 make it extremely challenging to safeguard the national network and information security. The number of regulatory rules on both the IP blacklist and whitelist have grown exponentially. When using IPsec-encrypted communication, the intermediate links do not have access to the communicated content. This renders security measures based on content identification obsolete, or at least less effective. Traceability after network and information security incidents need to be enhanced as a matter of urgency. In addition, the large-scale deployment of IPv6 technology in existing networks has not yet taken place. As such, the vulnerabilities and potential security risks in the protocol are yet to be thoroughly exposed, and further research in this area is needed. Products for network security, such as firewalls, security gateways, content filters, and intrusion detectors need to be developed to support IPv6.

4.3 China's IPv6 transition plan

The family of IPv6 protocols is becoming more complete, and the amount of IPv6 network element equipment is also gradually increasing. On this basis, IPv6-only networks can be set up. However, there is also a need to consider how IPv6 can be brought into IPv4 networks in actual application environments.

Research into IPv4/v6 integrated network techniques involve a wide range of contents, and includes all aspects of the network including network structure, function aggregation, network routing, domain name system, address allocation, service quality, typical applications, management functions and interfaces, and security requirements.

In summary, most research into IPv4/v6 integrated networking techniques includes the following:

1) The technical characteristics and scopes of application of the various transition strategies and network transition tools

Following years of research by the IETF's Next-Generation Network Transition Technical Task Force (NGTRANS) into transition techniques from IPv4 to IPv6 networks, numerous strategies (such as dual-stack, tunneling, and translation) and their corresponding transition tools have been proposed. These strategies and tools are applicable in different environments and under different conditions. Consequently, their effects are also different. Comprehensive comparison and comparative analysis of these strategies and tools can further clarify their scope of application and technical characteristics to determine their status and application environment in the networks.

2) IP 承载网络中引入IPv6 后的网络结构

The IP bearer network lies between the transmission network and the service network. It can be further divided into two layers: the IP transmission sublayer and the IP bearer control sublayer. When IPv6 technology is introduced into the IP transmission sublayer, the network structure and network logic structure of the IP bearer network are changed. These changes not only affect the service layer, but also impact the underlying transmission network. Through analyzing the network structure of the IP bearer network after IPv6 is introduced, the functions and related layers of each network layer in IP-based operating networks can be further clarified. This is of great significance in analyzing the interconnectivity and interoperability of networks, as well as their service provision methods.

3) Technical requirements imposed by telecommunication networks on IPv4/v6-integrated networks in different network environments

The application scope of IP technology in telecommunication networks is becoming wider. In terms of design concept, network architecture, modes of operation and maintenance, and service provision methods, telecommunication networks are very different from Internet networks. The typical network environment requirements that the IETF's IPv6 Network Interoperability Task Force (IPv6OPS) is investigating are specifically for the Internet. Although the research findings are somewhat useful when it comes to analyzing the typical network environment requirements of telecommunication networks, there are also limitations. Therefore, it is necessary to pay sufficient attention to research into the various typical network environments of telecommunication networks. Different network environments will impose different requirements on integrated IPv4/v6 networking technology. These typical network environments include: backbone networks, metropolitan area networks (including access, aggregation, and core), and customer premises networks.

4) Possible integrated networking solutions for different network environments

Several possible integrated networking solutions have been proposed for certain typical network environments. The technical characteristics, application scope, application outcomes, and limitations of these networking solutions were then analyzed. These networking solutions will soon become possible networking solutions of telecommunication networks. The outcomes of comparative analysis of these solutions will become the basis for selecting networking solutions for telecommunication networks.

5) Routing problems in IPv4/v6-integrated networks

When it comes to IPv4/IPv6-integrated network technology, routing problems are a key issue that must be closely investigated. The aspects of the issue include route selection, route reachability analysis, route leakage, routing loopback, route aggregation, and route reachability when different transition strategies are applied together.

Route leakage refers to the leakage of routes between IPv4 networks and IPv6 networks. As a result, the number of routes in areas (AS) with different routing policies increases, raising the complexity of routing management, reducing routing efficiency, and possibly even resulting in route loopback.

In order to improve the efficiency of network routing and reduce the size of the network routing table, one possible approach is to aggregate routes. Much emphasis has been placed on this aspect in the study of IPv4 routing policies. As the number of IPv6 addresses is larger, route aggregation is a necessity in IPv6 routing strategies. Therefore, sufficient attention was paid to this when the IPv6 protocol was first designed. In the IPv4/v6-integrated networking environment, route aggregation becomes more complex and requires in-depth analysis.

6) Domain name problems in IPv4/v6-integrated networking

Domain name resolution/reverse resolution is a type of service in an IP network, and also a basic function of an IP network. Many information retrieval services require DNS functions. In IPv4/v6-integrated networking environments, there are two main domain name issues that should be considered:

- The identification and resolution methods of IPv6 address-related domain names. Research in this area has already been stipulated in the RFC of the IETF;
- Domain name space/connectivity issues. This is technically challenging, especially the engineering aspect.

The presence of DNS servers that support different IP protocols can cause the DNS to split. The outcome is that some DNS servers cannot be reached from the root server by using the same IP protocol, hence some domain names cannot be resolved.

There are many possible solutions to this problem:

- Making all DNS servers in the integrated networking environment support dual-stack IP protocols;
- Making use of DNS proxy servers (which are actually dual-stack servers) to proxy some DNS queries so that the DNS space that supports IPv4 addresses is interoperable with that which supports IPv6 addresses;
- Formulating a corresponding DNS management policy to ensure that at least one DNS server in a DNS region is reachable via both IPv4 and IPv6. Even though these solutions are available, problems will still be encountered in the actual networking. For example, at present, host operating systems rarely support the sending of DNS messages in the form of IPv6 packets. This poses a challenge for network design.

7) Security in IPv4/v6-integrated networking

The security of IPv4 has faced much criticism. For this reason, the need for network security was emphasized during the design of the IPv6 protocol. A security header is included in the IPv6 packet header, which can be communicated based on AH or ESP, thus improving the security of the network to some extent. However, in an integrated networking environment, this problem becomes more complex. On the one hand, the security features of IPv6 cannot be fully realized in an integrated networking environment due to the existence of IPv4. On the other hand, the use of tunneling techniques in an integrated networking environment also introduces new security concerns. In addition, various comprehensive networking techniques require corresponding security analysis. The security of the IP bearer layer is only one aspect of network security, and includes multiple layers such as the service layer, service control layer, bearer layer, and support layer. When analyzing the security of the bearer layer from the perspective of the entire network, the problem becomes even more complex.

8) Strategies for allocating addresses in IPv4/v6-integrated networking

The IPv6 address allocation policy is specified by the relevant RFC from the IETF. Recently, progress has been made with regard to the IETF's address allocation proposal, such as the newly promulgated RFC 3177, while RFC 3587 replaces RFC 2374, and RFC 3513 replaces RFC 2373. Meanwhile, a detailed address allocation method has been determined by the relevant Internet management organization. What is discussed here is the way in which an operator allocates addresses in its networks, the guiding principle governing address allocation, and an analysis of address requirements related to specific networking techniques.

9) Interconnectivity and interoperability (compatibility) of different networking techniques in a network

At present, problems may arise when the various IPv4/IPv6 integrated networking techniques are used together in a network. For example, the interconnectivity and interoperability between networks built using these techniques is a major concern. Sometimes, a network that has been built using a certain networking technique cannot communicate with networks built using other networking techniques. There are many aspects to this, such as issues with address types, routing, domain names, and protocol translation. As such, careful analysis needs to be carried out to provide an important basis for evaluating and selecting the appropriate IPv4/v6-integrated networking technique.

4.3.1 The basic principles of network evolution

The evolution of operator networks to IPv6 is a long-term system-wide project involving user terminals, access networks, metropolitan area networks, core networks, information sources, and service supporting systems. Smooth evolution is necessary to protect IPv4 equipment in the existing networks, and to minimize the impact on both users and applications. As such, there is a need to adhere to the following basic principles:

1) Service-driven

The evolution of the next-generation Internet is primarily driven by services. On the one hand, there is a need to ensure that the development of existing services is not restricted by IPv4 addresses. On the other hand, the developmental needs of emerging services (such as the mobile Internet, cloud computing, the Internet of Things, and tri-network integration) need to be supported as well.

2) Smooth transition

The stable operation of existing networks must be ensured during the transition phase, and disruption to IPv4 users and services must be minimized. Transitional techniques and networking solutions should have minimal impact on both network architecture and system architecture so as to ensure a smooth transition.

3) Technological innovation

Different scenarios and requirements will arise at each stage of the transition, so attention must be paid to technological innovation and the implementation of new technologies. New IPv6 services should be introduced gradually, and IPv6 network traffic should be increased in order to improve the user experience.

4) Cost considerations

On the basis of meeting service needs for network evolution, transition strategies should also plan for the difficulty and cost of network deployment as far as possible. For sustainable development, the complexity of network deployment must be reduced, and a gradual approach must be used.

In formulating specific plans, it is also necessary to consider the problems caused by the incompatibility between the IPv4 and IPv6 protocols. It is necessary to come up with a transition mechanism to support the seamless coexistence of the two. The following guidelines should be followed:

- Ensure interoperability between IPv4 and IPv6 hosts, and support the interoperability of IPv4 services and IPv6 services on the premise that they do not affect each other;

- Ensure the normal functioning of existing IPv4 applications in the integrated networking environment;
- Avoid dependency among devices; device updates must be independent;
- The integrated networking process should be easily understood and implemented by network managers and end-users;
- Improve networking flexibility, and support the progressive upgrading of the network. Users are allowed to choose when and how to go about the transition;
- The quality of service of the network after integrated networking should not be visibly affected;
- The reliability and stability of the network after integrated networking should not be weaker;
- Network management should be strengthened after integrated networking;
- When designing an integrated networking solution, the long-term coexistence of IPv4 and IPv6 and the issue of smooth network transition need to be considered.

4.3.2 Transition

1) Operator network architecture

Currently, the three major operators in China are moving towards a flatter network architecture. The typical architecture is shown in Figure 4-3. The access networks (including wired and wireless access), metropolitan area networks, and backbone networks lie between the user terminals and the providers of applications and services.

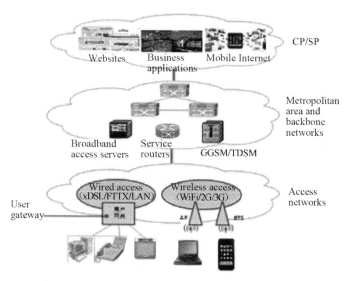

Figure 4-3: Typical architecture of network operators

2) Transition scenarios

The transition scenarios for operator network upgrades can be analyzed from the dimensions of users, networks, and service providers.

(1) Users

Fixed-network users can be divided into two categories according to their access-network environments. The first is connected to the operator's network through Layer 2 bridging devices (such as ADSL and FTTx), and the other through Layer 3 routing devices (such as CPE gateways with routing capabilities). For the former, support for IPv6 mainly depends on the operating system at the terminal, both to support the IPv6 protocol stack and to provide an IPv6 programming interface for the upper-layer applications. As for the latter, support for IPv6 depends on both the operating system at the terminal and the dual-stack capabilities of the CPE device. In the transitional phase, there are several ways in which the user terminal obtains the IP address through the network:

- Only the IPv4 address is obtained (the operating system at the terminal does not support IPv6);
- Both the public network IPv4 address and IPv6 address are obtained;
- Both the private network IPv4 address and IPv6 address are obtained;
- Only the IPv6 address is obtained (e.g. in closed Internet of Things application terminals).

The corresponding applications can be divided into those that support IPv4 only, those that support both IPv4 and IPv6 dual-stack, and those that support IPv6 only.

(2) Networks

Network evolution is a gradual process. In the transitional phase, multiple network forms will coexist, including single-stack IPv4 networks, dual-stack IPv4/IPv6 networks, and single-stack IPv6 networks. At present, the technical evolutionary pathways for both metropolitan area networks (MAN-s) and backbone networks are relatively well-defined. IP networks make use of the dual-stack approach, while MPLS networks use the 6PE technical pathway. After years of network expansion, equipment replacement, and upgrading, most devices on MAN-s and backbone networks can support both IPv4 and IPv6 stacks. As for access networks, due to the diversity of access techniques (such as ADSL, Ethernet, FTTx, and WLAN), equipment capabilities vary. This is therefore the most challenging aspect of upgrading the entire network, and as such is the focus.

For Layer 2 access networks, certain specific applications (such as multicast) require network devices to be able to listen to Layer 3 packets. Overall planning for Layer 3 access networks needs to be carried out alongside the smooth evolution of networks, and their deployment should be based on the dual-stack technique supplemented by the tunneling technique.

(3) Service providers

The upgrading and transformation of Internet service applications should also be a gradual process. It should be invisible to users and should not have a significant negative impact on their service experience. In the transition phase, applications will come in three forms: those that support IPv4, those that support IPv6, and those that support both. It must be guaranteed that IPv4 users can access existing IPv4 services without being affected, and that IPv6 users can access existing IPv4 services too. IPv4 users must be able to access new IPv6 services.

3) Analysis of typical scenarios

The transition of networks from IPv4 to IPv6 will be a long-term evolutionary process. There will be two main communication models during the transitional period.

(1) Communication between the same type of protocol

This means that both communicating parties have the same type of protocol (for example an IPv4 user accessing IPv4 services, and an IPv6 user accessing IPv6 services), and can traverse networks with different protocol types by making use of tunneling (such as 4-6-4 and 6-4-6). They can also traverse networks with the same protocol type (such as 4-4-4 and 6-6-6).

(2) Communication between different types of protocol

This means that the communicating parties have different types of protocol (for example an IPv4 user accessing IPv6 services, and an IPv6 user accessing IPv4 services). The problem of mutual access between IPv4 and IPv6 applications needs to be resolved.

4.3.3 Technical solutions for network transition

1. Overall architecture

For the sake of simplicity, operator networks can be divided into access networks, MAN-s, and backbone networks. Access networks can be further divided into bridging and routing types. For the evolution of the network, a combination of dual-stacks,

Table 4-1: Description of IPv4-to-IPv6 transition scenarios

Communication model	User	Network			Service	Scenario	Description
		Access Network	MAN (POP and above)	Backbone			
Same protocol type and network structure	IPv4	IPv4	Dual-stack	Dual-stack or 6PE/6VPE	IPv4	4-4-4	IPv4 user accessing IPv4 services using an IPv4 network
	IPv6	IPv6	Dual-stack	Dual-stack or 6PE/6VPE	IPv6	6-6-6	IPv6 user accessing IPv6 services using an IPv6 network
Same protocol type but different network structure	IPv4	IPv6	Dual-stack	Dual-stack or 6PE/6VPE	IPv4	4-6-4	IPv4 user accessing IPv6 services using an IPv6 network
	IPv6	IPv4	Dual-stack	Dual-stack or 6PE/6VPE	IPv6	6-4-6	IPv6 user accessing IPv4 services using an IPv4 network
Different protocol type but same network structure	IPv4	IPv4	Dual-stack	Dual-stack or 6PE/6VPE	IPv6	4-4-6	IPv4 user accessing IPv4 services using an IPv6 network
	IPv6	IPv6	Dual-stack	Dual-stack or 6PE/6VPE	IPv4	6-6-4	IPv6 user accessing IPv4 services using an IPv4 network
Different protocol type and network structure	IPv4	IPv6	Dual-stack	Dual-stack or 6PE/6VPE	IPv6	4-6-6	IPv4 user accessing IPv6 services using an IPv6 network
	IPv6	IPv4	Dual-stack	Dual-stack or 6PE/6VPE	IPv4	6-4-4	IPv6 user accessing IPv4 services using an IPv4 network

tunneling, and technical translation solutions can be used. In the evolution of access networks, one of the three following scenarios may arise: IPv4 access networks, IPv6 access networks, and IPv4/IPv6 dual-stack access networks. Either the 6over4 or 4over6 solution can be considered. The two endpoints of a tunnel are the home gateway on the user side (or terminals, by loading software plug-ins), and the equipment (such as the CGN and gateways) on the network side. As for backbone networks and MAN-s, the dual-stack solution is used for the IP-bearer type while the 6PE solution is used for the MPLS-bearer type.

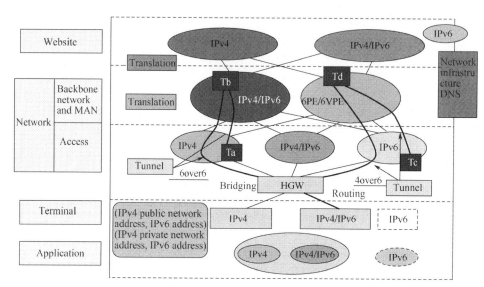

Figure 4-4: Technical solutions for the transition to IPv6

2. Backbone networks

(1) Based on IP technology

Deploying the most mature IPv6 solution in Native IP backbone networks means the use of the dual-stack technique, which requires an upgrade to existing network equipment.

(2) Based on MPLS technology

The simplest way to deploy IPv6 in MPLS backbone networks is to maintain single-stack P (core) IPv4 routers with dual-stack PE routers at the network periphery, so that IPv6 traffic is transmitted via the IPv4-based MPLS label paths.

(3) Technical characteristics

- In an IPv4 network, the IPv4 IGP protocol (such as IS-IS) is still used between the PE and P routers to establish the routing relationship between them, and an IPv4 label distribution protocol (such as LDP or RSVP-TE) is still used to establish an LSP (label-switched path) between two PE and PE. Therefore, IPv6 networks are invisible to both P and PE routers.
- The MP-BGP multi-protocol extension attribute needs to be used to propagate routing information for IPv6 networks.
- Neither 6PE nor 6VPE routers are dedicated devices. While providing tunneling or VPN services for IPv6 networks, they can also provide MPLS VPN services to ordinary IPv4 users on other interfaces or sub-interfaces.

(4) Implementation plan

The following methods can be used to provide IPv6 services on MPLS backbone networks:

- Deployment of IPv6 in the PE routers at the periphery of the MPLS backbone network, providing IPv6 Internet access in existing routers that have IPv4 Internet access and are located at the periphery of the network;
- In VPN and Internet access services, an IPv4 Label-Switch Path (LSP) is used to carry IPv6. By adopting 6PE or 6VPE, it is possible to avoid modifying the configuration of the existing network in the P routers at the core of the network. These include the existing interior gateway protocol (IGP), label distribution protocol (LDP), and existing addressing arrangement.
- During deployment, special attention must be paid to the issues with ICMPv6 and IPv6 MTU. As the P routers at the core of the network do not support IPv6, ICMPv6 packets will be discarded. As such, the path MTU discovery mechanism must be implemented at the periphery of the network.

3. MAN-s + access networks

MAN-s consists primarily of Layer 3 routing equipment (CR-s and core routers) and service control equipment (SR-s, service routers, BRAS-s, and broadband access servers). They forward IP packets and carry out service diversions. Access networks mainly consist of Layer 2 access devices (DSLAM-s, switches, and OLT-s/ONU-s), and are primarily used for providing access to IP packets.

For MAN-s, a dual-stack technical pathway should be used above the point of presence (POP), and a corresponding gateway device should be deployed to terminate the tunnel and complete protocol conversion as required. The transitional technical

Table 4-2: A comparison of native IPv6 and 6PE/6VPE

	Native IPv6	*6PE/6VPE*
Implementation costs	All equipment needs to be upgraded to support IPv6, so implementation costs are high	Only PE devices and not P devices need to be upgraded to support IPv6, so implementation costs are low
Protocol deployment	All routers make use of IPv6 IGP/BGP	The MPLS core is unchanged while MP-BGP is deployed between PE-s
Scalability	No restriction	No restriction
Maintenance costs	All nodes need to maintain the newly-introduced IPv6 protocol and routes	IPv6 is a new MPLS service and does not affect the maintenance of the existing networks. The scope of maintenance is limited to the PE
Service support	Unicast/multicast	Supports VPN services, but multicast is still immature

solution for access networks should be based on the dual-stack technique, combined with tunneling and translation mechanisms. The transition plan will vary according to the application scenario.

- When there is no obvious change in content and applications, operators prefer the CGN technique, and use dual-stack private networks with NAT444;
- When there is no obvious change in content and applications, operators tend to prefer IPv6. DS-Lite is used to provide users of private IPv4 networks with access, and NAT44 is deployed to convert private IPv4 networks into public IPv4 networks;
- When most content and applications have been converted to support dual-stack, the dual-stack technique is used;
- When the user switches to single-stack IPv6, 6RD is used to provide access for IPv6 users, and the NAT64/IVI translation mechanism enables IPv6 users to access the service resources of IPv4 networks;
- When most of the content and applications have been converted to IPv6, DS-Lite is used to provide access for public network IPv4 users, and the IVI translation mechanism enables IPv4 users to access the service resources of IPv6 networks.

For large Layer 2 access networks, the access devices transparently transmit the forwarded packets. However, the devices should be able to identify and distinguish between IPv6 and IPv4 packets, and perform VLAN tagging, QoS, and packet-filtering according to the IPv6 protocol. If dual-stacks are supported, link-layer data protocols such as PPPoE and IPoE can be used to carry the IPv6 protocol directly.

Table 4-3: Transition scenarios and their corresponding technical solutions

Communication model	User	Network			Service	Scenario	Technical solution
		Access Network	MAN (POP and above)	Backbone			
Same protocol type and network structure	IPv4	IPv4	Dual-stack	Dual-stack or 6PE/6VPE	IPv4	4-4-4	Dual-stack private networks + NAT444
	IPv6	IPv6	Dual-stack	Dual-stack or 6PE/6VPE	IPv6	6-6-6	
Same protocol type but different network structure	IPv4	IPv6	Dual-stack	Dual-stack or 6PE/6VPE	IPv4	4-6-4	DS-Lite
	IPv6	IPv4	Dual-stack	Dual-stack or 6PE/6VPE	IPv6	6-4-6	6RD
Different protocol type but same network structure	IPv4	IPv4	Dual-stack	Dual-stack or 6PE/6VPE	IPv6	4-4-6	This scenario will only appear in the long term when IPv6 networks and applications are mainstream
	IPv6	IPv6	Dual-stack	Dual-stack or 6PE/6VPE	IPv4	6-6-4	NAT64/IVI
Different protocol type and network structure	IPv4	IPv6	Dual-stack	Dual-stack or 6PE/6VPE	IPv6	4-6-6	This scenario will only appear in the long term when IPv6 networks and applications are mainstream
	IPv6	IPv4	Dual-stack	Dual-stack or 6PE/6VPE	IPv4	6-4-4	6RD + NAT64/IVI

Explanation:

- Dual-stack private network + NAT444: For bridging access networks, only central office equipment such as BRAS-s and OLT-s needs to be upgraded to support dual-stacks, so that the terminal can be allocated both an IPv4 and an IPv6 address simultaneously. For routing access networks, both the user's home gateway devices and central office equipment (such as BRAS-s and OLT-s) need to be upgraded to support dual-stacks and NAT44. At the same time, planning needs to be carried out to ensure that there is no conflict between the private network address space at the user side and private network address space at the operator's network access side. The user terminal is able to obtain a private IPv4 network address and an IPv6 address.
- DS-Lite: This solution is applicable to routing access networks, without needing to consider the installation of plug-ins to implement client functions at the user terminal. The user's home gateway devices must be upgraded to support dual-stacks and B4 functions, and central office equipment must support dual-stacks, AFTR, and NAT44 translation functions. The user terminal obtains a private IPv4 network address and an IPv6 address.
- 6RD: This solution is applicable to routing access networks (without considering the installation of plug-ins to implement client functions at the user terminal). The user's home gateway devices must be upgraded to support dual-stacks and 6RD client functions, and the central office equipment must support dual-stacks, and 6RD gateway functions. The user terminal obtains an IPv6 address and the user's home gateway is allocated an IPv4 public network address.
- Stateful and interoperable NAT64: It allows IPv6 users to access IPv4 services (stateful), and NAT64 translation devices can be deployed at IDC exits or in MAN-s as required. At the same time, DNS64 devices are deployed accordingly.
- IVI stateless and interoperable solution: It allows IPv6 users to access IPv4 services, or IPv4 users to access IPv6 services (stateless). IVI translation devices can be deployed at IDC exits or in MAN-s as required. The DNS also needs to be modified and upgraded.

4) Domain Name System (DNS)

In the transition from IPv4 to IPv6, the difference between the DNS record format in IPv4 and IPv6 means that the DNS service that forms the Internet's basic framework needs to be upgraded and modified. At present, two different technical evolutionary solutions can be considered: one is to upgrade to dual-stacks, and the other is based on translation mechanisms.

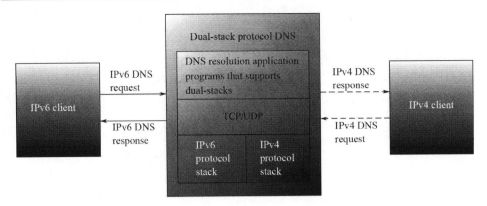

Figure 4-5: The dual-stack network + dual-stack DNS model

The DNS is upgraded to support dual-stacks so that access by an IPv6 user returns a record of "AAAA", while access by an IPv4 user returns a record of "A".

Another option is to use translation mechanisms such as DNS64. Generally, NAT64 and DNS64 work synergistically, and the application scenario is 6-6-4. The basic principle is that when an IPv6-only user accesses a single-protocol IPv4 stack server, the IPv4 address of the server is prefix composed via a DNS64 server (with the specific address prefix 64:FF9B::/96). Traffic with this specific prefix network segment will be routed and forwarded to the NAT64 router to implement IPv6 and IPv4 addresses and protocol conversion.

The DNS64 complements the NAT64 by synthesizing the A records (IPv4 addresses) in the DNS query information into the AAAA records (IPv6 addresses). The synthesized AAAA record is then returned to the user at the IPv6 side (see Figure 4-5).

5) Typical transition solutions

Due to the variety of Internet application scenarios, there is no unified model for technical solutions for the evolution of IPv6. For operators, IPv6 is the fundamental solution for the issue of IPv4 address shortage, and is also an evolutionary process. When selecting a transition technique, the following factors must be taken into consideration: existing network investments must be protected, costs must be reasonable, the difficulty of network transformation should be moderate at most, existing services should not be compromised, and the user experience should be good. The key to the selection and adoption of an evolutionary IPv6 technique is often more than just technical considerations. The greater challenge is the impact of Internet development and technical transformation on service and business models. Therefore, to formulate a strategy for smooth network evolution, the various service application scenarios and future network development needs must all be considered, as well as the above-mentioned factors. Transition techniques should be used in combination.

At present, the main technical solutions for network evolution are private network dual-stack + NAT444, DS-Lite + NAT44, 6RD, and the interoperable IPv6 and IPv4 solutions of NAT64/DNS64 and IVI.

(1) Solution 1: private network dual-stack + NAT444

NAT 444 is a two-level NAT, i.e. NAT44 at the user-side HG (home gateway) and NAT44 at the operator side (LSN devices) for two-level address translation. It thereby forms three address spaces: private addresses on the user side, private addresses on the operator side, and public network addresses. This solution requires both the user terminal and the home gateway to support dual-stacks. They are then allocated a private network IPv4 address (i.e. a different address space) and an IPv6 address. IPv6 service traffic is forwarded according to Native IPv6, while IPv4 traffic undergoes two-level NAT translation. The operator's CGN devices perform the translation to public network IPv4 addresses, as shown in Figure 4-6.

The NAT444 solution can increase the multiplexing rate of IPv4 addresses, alleviating the problem of address exhaustion. In addition, deployment is easy, since only home gateways need to be upgraded to support dual-stacks on the user side; on the network side, all that is required is for CGN devices to be added at the aggregation layer or core layer. In terms of user perception, maturity of technique, and ease of deployment, NAT444 is currently the preferred solution. However, adopting it will increase the difficulty of implementing P2P-type services and address tracing, and also means that end-to-end transparency cannot be achieved.

(2) Solution 2: DS-Lite + NAT44

DS-Lite is a typical IPv6 + 4over6 tunneling solution. The DS-Lite CGN equipment is deployed in MAN-s. The broadband access server connects to the core router through IPv6. The user's CPE supports DS-Lite, and an IPv4-over-IPv6 tunnel is established between the CPE and the DS-Lite CGN. The broadband access server assigns an IPv6 address prefix to the user's CPE, and the IPv4 address of the user's host is assigned an IPv4 private address by the CPE. The user's IPv6 traffic flows directly to the core router through the broadband access server. After the user's IPv4 traffic reaches the CPE, it passes through the IPv4-over-IPv6 tunnel to the DS-Lite CGN. At the same time, the CGN has the NAT44 function so that the user's IPv4 data is de-capsulated after tunneling, and NAT44 address translation is carried out once more. Finally, the data is sent to the IPv4 backbone network, as shown in Figure 4-7.

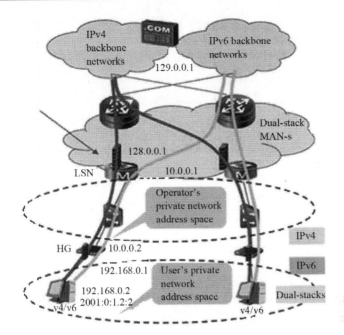

Figure 4-6: Private network dual-stack + NAT444

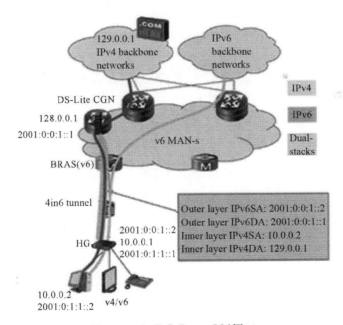

Figure 4-7: DS-Lite + NAT44

This solution requires the CPE terminal to support dual-stacks and be allocated a private IPv4 network address as well as an IPv6 address.

One of the advantages of adopting the DS-Lite solution is to relieve pressure on the broadband access server equipment so that only the IPv6 protocol stack is being operated. This makes it suitable for application scenarios in which IPv6 is dominant.

(3) Solution 3: 6RD

6RD is a typical IPv4 + 6over4 tunneling solution. Similar solutions include IPv6-over-L2TP (L2TP tunnels are used to provide IPv6 users with remote access). 6RD is a way to rapidly introduce IPv6 into IPv4-based networks. Equipment such as broadband access servers of existing MAN-s do not need to be upgraded to support IPv6. 6RD gateways are deployed in clusters in the MAN-s to establish IPv6 links with the backbone networks. In addition, the user's CPE needs to support 6RD. When the user requires IPv6 access, an IPv6-over-IPv4 tunnel is established between the CPE and the 6RD gateway. IPv6 traffic is forwarded to the 6RD gateway through the tunnel. Meanwhile, the user's IPv4 access is carried out using the original path. This solution is suitable for the early stages of IPv6 development, in which the bulk of services offered by operators are based on IPv4, and there is only a handful of IPv6 users.

The 6RD solution is shown in Figure 4-8.

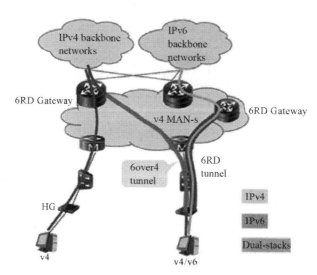

Figure 4-8: 6RD

Table 4-4: A comparison of the three transition solutions

	Private network Dual-stack +NAT444	DS-Lite+NAT44	6RD
Maturity of technique	Standardized	Standardized	Standardized
Commercial maturity	Large-scale deployment of NAT44 in existing networks; product is mature	Large-scale deployment not yet carried out; capabilities of equipment remain to be tested	Large-scale deployment carried out by some European operators
Impact on existing networks	Minor	Significant; central office needs to have DS-Lite tunnels and translation (44) functions	Significant; central office needs to have 6RD tunnels and translation (64) functions
CPE requirements(s) at client side	Translation of private network address space	To support DS-Lite tunnels	To support 6RD tunnels
IPv4 address requirement(s) at client side	Private network IPv4 addresses	Private network IPv4 addresses	Public network IPv4 addresses
Facilitates migration of traffic to IPv6?	No; traffic is still carried by IPv4 networks	Yes; traffic (access networks) carried by IPv6 networks	No; traffic is still carried by IPv4 networks
Main problems	Two-level NAT increases service complexity, leading to complex networks; address-tracing is difficult	New bottleneck in central office equipment (scalability issues), large-scale upgrading of CPE home gateways required	Large-scale upgrading of CPE home gateways required; applications need to support IPv6

(4) Interoperable solution 1: stateful translation technique – NAT64+DNS64

NAT64+DNS64 is a stateful translation solution for IPv6 users accessing IPv4 network service resources. With this solution, all applications, access devices, and networks on the client-side support IPv6. Users employ IPv6 services and content directly but require stateful address translation (NAT44) gateways to access IPv4 services and content. The DNS64 complements the NAT64 by synthesizing the A records (IPv4 addresses) in the DNS query information into the AAAA records (IPv6 addresses). The synthesized AAAA record is then returned to the user at the IPv6 side. The user's service traffic is then routed to the NAT64 gateway device based on this destination address. Address and protocol translation of both the destination address and source address are then carried out in the device before being routed to the destination server in the form of an IPv4 packet, as shown in Figure 4-9.

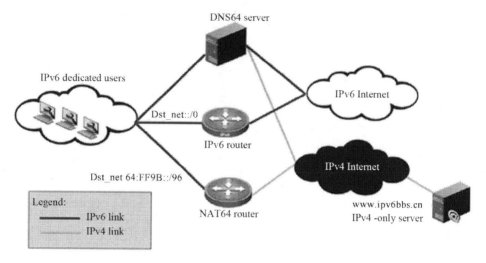

Figure 4-9: 6RD

(5) Interoperable solution 2: stateless translation – IVI

IVI is a stateless translation solution for IPv6 users accessing IPv4 network service resources and IPv4 users accessing IPv6 network service resources. This solution is essentially about constructing a specific IPv6 address segment by using an existing IPv4 address segment. An explicit and specific mapping relationship is formed by

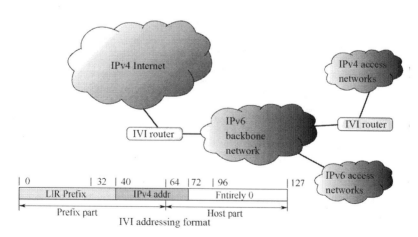

Figure 4-10: Architecture of the IVI solution

embedding the IPv4 address in the IPv6 address segment. IVI has two main functions. The first is address-mapping, in which standard rules are used to implement one-to-one mapping between IPv4 addresses and IPv6 addresses for address translation. The second is protocol translation, in which mutual translation of the relevant fields is carried out between the IPv4/ICMPv4 protocol and the IPv6/ICMPv6 protocol according to standard regulations. At the same time, the relevant fields of the TCP/UDP protocol are updated to complete the translation of the data packet.

Table 4-5: Comparison of two types of interoperable transition solutions

	NAT64	*IVI*
Maturity of technique	Standardized	Standardized
Commercial maturity	Large-scale deployment not yet carried out; maturity of products remains to be tested	Deployed in CERNET2
Address format requirements	Specific address prefix 64:ff9b::/96	Specific format, IPv4 address embedded in IPv6 address
Service access	Only unidirectional access supported, i.e. IPv6 users accessing IPv4 services	Bidirectional access supported, i.e. IPv6 users accessing IPv4 services, and IPv4 users accessing IPv6 services
Impact on the DNS	DNS upgraded to support DNS64	DNS upgraded to support IVI
Impact on existing networks	Minor. NAT64/DNS64 devices need to be deployed at nodes that require IPv4/IPv6 interoperability	Minor. IVI gateways, and a DNS that supports IVI functions need to be deployed at nodes that require IPv4/IPv6 interoperability

Notes

1. Source of data: The *29th Statistical Report on Internet Development in China* published by the China Internet Network Information Centre (CNNIC).
2. In the next five years, China's IP address demand will be 34.5 billion, including 1 billion for mobile Internet, an estimated 10 billion for the Internet of things, and another 500 million for fixed Internet. Estimation based on the assumption of IP address utilization rate of 33%.
3. Source: http://resources.potaroo.net/iso3166/v6cc.html.
4. Source: http://bgp.potaroo.net/v6/as2.0/index.html.
5. Source: http://www.caida.org/research/topology/as_core_network/historical.xml.
6. IPv6 Forum. IPv6-Ready certification comprises two phases: Phase 1 tests conformity to and interoperability with the IPv6 core protocol; Phase 2 includes optional protocol testing.

7. Operation systems including those in PCs and mobile phones.

8. Including those in mobile phones and network equipment.

9. According to preliminary research, the PXA920 TD chip supports the IPv6 stack, while the PXA1802 LTE chip supports the IPv4/IPv6 dual-stack.

10. The first level refers to the translation of the user's IPv4 private network address to the operator's IPv4 private network address; the second level refers to the translation of the operators' IPv4 private network address to the IPv4 public network address.

11. At present, dIVI is proposed as a solution. Multiplexing of IPv4 public network addresses can be realized through ports.2. Network structure of IP bearer network after introducing IPv6.

Chapter 5

Future Networks: Key Problems and Current Research

Chapter Highlights:
- *Architecture of future networks*
- *Service support capabilities of future networks*
- *External capabilities of future networks*
- *Current state of research and trends in new network architecture*
- *Testing platforms for future networks*

Overview

IPv6 can solve the current IP address shortage, but prominent unsolved issues still restrict the development of existing networks, such as security, scalability, and mobility. As such, some academics and industry experts have started using a "clean slate" approach, making use of revolutionary technological pathways to enhance the innovation of network architecture and basic protocols. This is done in a bid to determine the technical direction for the medium- to long-term development of the next-generation Internet, and has resulted in a new wave of enthusiasm for technical research into the networks of the future.

5.1 Network Architecture of Future Networks

5.1.1 Key problems in network architecture: naming, addressing, routing, and resource management

Throughout the development of data communications, naming and addressing have never been major concerns. This is especially so if the network architecture is simple enough and the scope very limited. Most of the early networks were point-to-point or multi-station line networks, so addressing was done using simple enumeration. Similarly, addressing was also not a problem in large SNA networks because SNA is hierarchical. This means that there is only one path from the terminal to the host, so enumerating the levels of the hierarchical structure is sufficient. In fact, even for a distributed network with multiple paths (such as the early ARPANET and even the early Internet), addressing can be done through enumeration. However, as the network structure grows more complex, naming and addressing become problems that must be tackled. Together with routing and resource management, they are the key network architecture problems of the Internet.

Since the early days of the Internet, there have been many theoretical studies on these issues. One of the most widely accepted ideas was proposed in a paper published in 1978 by John F. Shoch entitled Inter-Network Naming, Addressing, and Routing. The article was circulated within the ARPANET community for more than a year before it was published. The author believes that there are three important concepts in computer communications: the name of the position-independent application, which represents what we are looking for; the logical address with positional information, which indicates where it is; and routing options, which represent how to get there. In addition, there is a mapping relationship between the logical addresses and the underlying physical addresses. In the context of the Internet today, what corresponds to these concepts? The URI (URL or URN) of the application layer is responsible for naming. The IP address at the network layer corresponds to the logical address. Routing is mainly completed in the network layer, and the MAC address corresponds to the physical address, as shown in Figure 5-1.

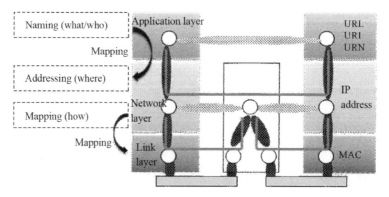

Figure 5-1: The ideal architecture of computer communications mapped to the existing Internet architecture

Careful analysis of the correspondence would lead to the discovery of problems in the core architecture of the present Internet, such as addressing, naming, routing, and resource management.

1) The lack of a naming mechanism at the application layer

In the structural mapping shown in Figure 5-1, URL, URI, and URN correspond to the naming mechanism of the application layer. However, as resource identifiers, URI and URN are closely related to location, so it is more appropriate to regard them as application layer addresses. Therefore, there is actually no position-independent naming mechanism at the application layer.

2) The network-layer IP address erroneously assumes double semantics

In the present Internet architecture, the IP address serves as both a location identifier and an identity identifier, which involves dual semantics. From the perspective of the transport layer and the application layer, the IP address represents the identity information, and is used to identify an end-to-end connection. From the perspective of the network layer, the IP address represents the location information that is used for routing and addressing in the network. In the early days of the Internet's development, the use of an IP address as a terminal's identity indicated its global uniqueness. This did away with the need to introduce new namespaces, thus simplifying the design and implementation of transport-layer protocols. However, an IP address is actually the addressing of an interface, and thus changes with location. Therefore, when used as a name, it is not position-independent. When used as an address, the IP address is more a reflection of information related to the operator because of the way in which IP addresses are globally applied for and allocated at present. As such, IP addresses do not fully convey geographical information, unlike in telephony. They also cannot be strictly stratified, unlike logical addresses in operating systems.

3) Scalability – a serious problem in Internet routing

The root cause of the routing scalability problem is the double semantics of IP addresses within the present Internet architecture. An IP address represents both the host's identity (representing the endpoint of a session at the transport layer) and the host's location (used for routing data packets in the routing system). According to Rekhter's Law, in order to ensure the scalability of the routing system, the allocation of IP addresses should adapt to the network topology. However, because an IP address represents the host's identity at the same time, address allocation is often based on organizational structure rather than topological structure. As such, it is relatively stable, and cannot be dynamically adjusted according to changes in the network topology. These two roles have conflicting aims, making it difficult for a single IP address namespace to meet the requirements of both roles at the same time, which results in routing scalability issues.

4) Contradictions in Internet resource management

At present, the Internet has two types of basic network resources. The first comprises domain names used for basic Internet applications such as Web-browsing, email, and virtual network communities. The other comprises IP addresses used for host and location identification. Currently, the management of the domain name system is being controlled by developed countries. ICANN (the Internet Corporation for Assigned Names and Numbers) – the main assignment organization for Internet domain names and numbers – is located in the United States. All 13 root servers are also located in developed countries. As for IP addresses, their allocation is too simplistic. The IP address is more a reflection of information related to the operator because of the way in which IP addresses are globally applied for and allocated at present. As such, IP addresses do not fully convey geographical information, unlike in telephony. They also cannot be strictly stratified, unlike logical addresses in operating systems. Objectively speaking, this has distorted the semantics of IP addresses.

To sum up, the four aspects of naming, addressing, routing, and resource management that constitute the core architecture of the Internet currently face problems such as an incomplete naming mechanism at the application layer, addresses with double semantics at the network layer, serious routing scalability issues, and contradictions in Internet resource management. The issues with the core architecture of the Internet are an important cause of many other problems. The lack of a naming mechanism makes the Internet inherently inadequate when it comes to mobility support capabilities, although there are some solutions such as MobilIP and dynamic DNS. To tackle the mobility issue, implementation is still challenging. In addition, the emergence of "multi-homed hosts" makes the problem more complex. These problems have become a bottleneck restricting the development of the Internet.

5.1.2 Establishing a unified naming and mapping mechanism to overcome naming problems

In future networks, network identification, and user identification must be separated. This is an inevitable requirement for the separation of networks and services. However, the relationship between service identifiers, network identifiers, and user identifiers must also be taken into consideration. Future networks will support various services.

Therefore, to streamline operations and facilitate usage, it would be best if the user only needs to remember one memorable user identifier. Service identifiers and network identifiers are maintained and managed by operators, so that features similar to ENUM and "one pass" can be provided. To summarize, the most important direction for the addressing and naming system of future networks is the combination of unified naming with the rapid mapping mechanisms of network resources. This marks a return to the original ideas of providing users with a location-independent user identifier at the application layer; of network layer addresses being location identifiers; and of naming and addressing functions to be clearly separated and defined, to establish a rapid mapping mechanism between the two. There are three technical approaches in this general direction:

1) A naming mechanism at the application layer based on the existing Internet mapping mechanism

In the application layer, each entity is assigned a unique identifiable resource identifier. The syntax of the resource identifier defines the character set that is used to name the network resources and constitutes the rules; the syntax rules of the character set determine the range and size of the Internet resource namespace. A collection of all Internet resource names constitutes a namespace. Existing DNS can be used to establish mapping between the resource names and URI-s, so that an addressing system based on URI can be created. Notable solutions in this area include the XRI and UDDI proposed by the W3C.

The application-layer naming system based on the existing Internet mapping mechanism is shown in Figure 5-2.

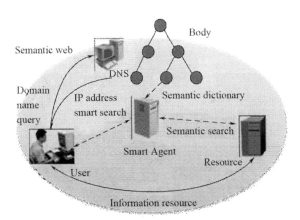

Figure 5-2: An application-layer naming system based on the existing Internet mapping mechanism

2) A naming system based on distributed mapping

Content identifiers are used for application layer naming. The names of applications or nodes do not vary with location, thus solving mobility problems at the application layer. At the same time, a distributed query system is used to directly implement efficient mapping between application-layer naming and network-layer node addresses. This approach has good scalability. A notable solution in this area is the I3 network proposed by Berkeley.

The naming system based on distributed mapping is shown in Figure 5-3.

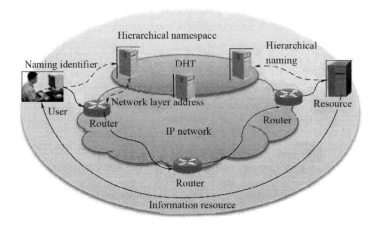

Figure 5-3: A naming system based on distributed mapping

3) Name-based content routing

The application layer and the network layer are merged to establish a hierarchical namespace. For naming purposes, the content is mapped as content identifiers, and is used directly in name-based content routing. A notable solution in this area is the NDN (Named Data Networking) network.

Name-based content routing and transmission control is shown in Figure 5-4.

5.1.3 Establishing a highly scalable addressing system with clear semantics to overcome addressing problems

At present, there are problems with both the syntax and the semantics of IP addresses. Syntax refers to the syntactical structure of the address, including the header design and the length of the address. The semantics of the IP address refer to its roles, and to the address-allocation mechanism.

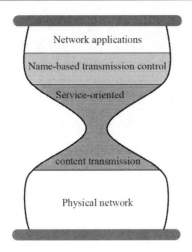

Figure 5-4: Name-based content routing and transmission control

From a syntactical point of view, the current IPv4 address is only 32 bits long, and its capacity is unable to support the development of the Internet. At present, all of the IPv4 addresses in the IANA address pool have been assigned, and the remainder of the RIR's inventory is expected to last two years at most. Address scalability is currently one of the most pressing issues in the development of the Internet.

From a semantical point of view, the problem of dual semantics for IP addresses needs to be resolved. An IP address contains identity information, and identifies an end-to-end connection. It also contains location information, and is used for routing and addressing in the network. This ambiguity has led to problems in routing scalability, mobility, and multi-homing, and is a bottleneck restricting the Internet's development.

There are three main directions for solving these problems:

1) Remedial measures to resolve the issue of address shortage without changing the syntax and semantics of the IP address

There are three main techniques in this gradual approach. The first would be to improve the usage efficiency of existing addresses, represented by classless inter-domain routing (CIDR), dynamically-assigned IP addresses, and variable-length subnet masks (VLSM). Second is address translation, such as the translation of private addresses/ network addresses (NAT). Third is address embedding, such as IP in IP tunneling and DS Lite. These remedial measures can alleviate the shortage of addresses in the short term, but do not fundamentally solve the problem. The widespread usage of address-multiplexing techniques will accelerate the deterioration of the Internet's operating environment and development worldwide. It will also cause the network to become more complex more quickly, and will lead to cost increase in business innovation, deployment, and operation. Not least, it will pose new challenges for security issues such as traceability. Therefore, to completely solve the problem of address scalability, the grammatical structure of IP addresses must be changed.

2) Reshaping the semantics and grammar of network-layer addresses to resolve problems such as scalability, semantic ambiguity, and derived routing scalability.

A revolutionary technique would be to design addresses with hierarchical structures. The address structure could be redesigned so that network-layer addresses had hierarchical semantics (according to the hierarchy of the network topology or geographical area) and would include routing information such as potential node energy. When designing the structure of the address, the length could be increased to solve the problem of scalability; geographical information could be included in the structure to clarify the location functions of the network-layer address, to resolve problems such as semantic ambiguity; address fields could be switched to replace routing, to overcome problems such as routing scalability. This revolutionary approach of changing both the semantics and grammar would require a complete overhaul of the current address system. The process would involve the transformation of all aspects of the network. It is currently still in the research and testing stage, and there is still quite a long way to go before a large-scale roll out and public acceptance.

3) Preserving the semantics of the IP address but changing the syntax to completely resolve the address shortage issue, leaving room for innovation in terms of ease of usage, multicasting, and service quality.

IPv6 is representative of this blended approach. Compared to solutions using the gradual approach, IPv6 has an almost unlimited address space. It can guarantee the end-to-end transparency of IP networks and simplify the network structure, as well as completely solving the problem of address scalability. Compared to the revolutionary approach, IPv6 is the only technology in the world that is sufficiently mature for large-scale commercial deployment in the industry. As such, IPv6 deployment is the best and most inevitable option at this phase of the Internet's evolution. However, even though IPv6 deals with the issue of address shortage, there is no substantial improvement over IPv4 in other aspects such as routing scalability, security, and assurance of service quality. Therefore, even after IPv6 deployment, there will still be a technical evolutionary phase in which breakthroughs are needed in multiple technical areas.

5.1.4 Moving from a flat routing mechanism to a hierarchical one to overcome routing issues

The most prominent problem with the existing routing architecture is scalability. According to an analysis report of the Border Gateway Protocol (BGP) routing table data provided by APNIC (Asia-Pacific Network Information Center), the number of Internet routes is increasing rapidly. As of December 2010, there were 336,364 IPv4 BGP routes, as shown in Figure 5-5. A bold prediction by experts in the industry claims that there will be two million entries in the Internet routing table by 2020.

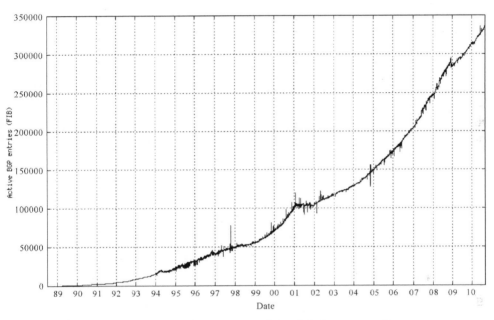

Figure 5-5: Rapid expansion of the global routing table

There are various reasons for routing scalability problems, as shown in Figure 5-6. These include the rapid development of the network/user scale, the widespread use of operator-independent addresses, the long-term coexistence of IPv4 and IPv6, the ever-increasing IPv4 address transactions, the widespread use of network multihoming and traffic engineering, and the dual semantics of IP addresses. Amongst them, the most fundamental reason is the latter.

The rapid expansion of the routing table will have severe consequences for the Internet's development. They include a slowing down in the convergence of inter-domain routing leading to degraded network stability; an increase in R&D costs for manufacturers and investment costs for operators; an increase in energy consumption of equipment that is inconsistent with the trends for energy conservation and reduction of carbon emissions.

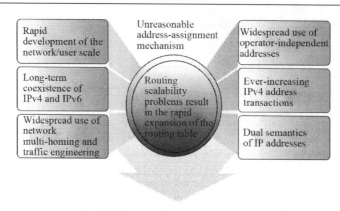

Figure 5-6: Causes and consequences of routing scalability problems

In order to solve problems such as routing scalability, the Internet industry has begun research into routing architecture, and has come up with many ideas and solutions. The most important solutions can be categorized into the gradual approach and the revolutionary approach according to the technical principles behind them.

The gradual approach does not change the semantics of the IP address. Instead, some remedial measures are taken to alleviate the routing scalability problem. The specific techniques are described below.

1) Separation of edge-user networks from operator networks (Map/Encaps)

The idea is to separate the ISP network (RLOC or Routing Locator) from the edge-user network (EID, Endpoint ID). Only the globally routable address of the ISP network is preserved in the core routers. This means that large numbers of edge-user addresses will be removed from the core router, greatly reducing the number of routes. Routing in local networks is based on local location identifiers, i.e. unique identification and local addressing. When data packets enter the core networks, RLOC must be added by the encoder. Routing in the core networks is based on the RLOC. Once the data packets enter the ISP networks identified by the RLOC, the final transfer is taken over and completed by local routing. The preliminary plan is to allocate RLOC only for ISPs with important interconnection links, of which there are fewer than 10,000 worldwide. Moreover, the growth rate of this figure is much lower than that of the routing table. Notable examples of this technique are Cisco's LISP (Locator Identifier

Split Protocol) and Huawei's VA (Virtual Aggregation), which support incremental deployment.

Although this type of solutions helps to alleviate the problem of routing scalability, it also gives rise to new problems, such as the need to add a tunnel encapsulation to all IP packets; the need for mapping between the Identifier and the Locater; the generation, distribution, and queries of huge mapping databases; possible timeouts for the first data packet; and deployment of new protocols. These problems impose a great burden on the network, and are costly to solve.

2) The separation of identity and location (ID/locator)

This approach advocates the complete separation of ID and locator. The address is used only to identify the location and to provide global routing. A new namespace is created for the ID. When a host sends a data packet, the source and destination locators are added. In the subsequent transmission through the network, only the locators are used for routing. When the mapping between the ID and locator changes, the host advertises to the distributed mapping database. The host requests and caches the required identifier mapping from the mapping database. If a new user identifier is defined in existing IP networks, the IP address can be deployed according to the network topology, facilitating CIDR address aggregation, and thus thoroughly solving the problem of routing scalability. Notable examples of this technique are Ericsson's Host Identity Protocol (HIP) and Huawei's Hierarchical Routing Architecture (HRA).

The separation of ID and locator enables the complete decoupling of the dual semantics of an IP address, thus solving problems such as mobility and network multi-homing, enhancing security, and allowing scalability of the routing system. However, such techniques also involve modifying the hosts. They are relatively difficult to deploy, and cannot support effective traffic engineering and multicasting. In addition, the overheads of the protocol are large, especially for mobile application scenarios in which bandwidth resources are limited.

3) Aggregation of geographical locations

Aggregating location identifiers based on physical regions allows the aggregation of routing information based on direction. Since there is no need to flood reachability information (which contains location identification), this solution can greatly reduce the number of entries in the routing table. The problem is that it does not conform to the Internet business model, and regional interconnection points need to be established.

4) Forwarding table compression

Internet routing is completely unchanged and continues to use the BGP protocol. However, when generating an FIB table, an algorithm is used to compress the entries to be forwarded. Even though this solution does not improve the scalability of the control plane, it greatly improves the scalability of the forwarding plane. However, the issue of RIB scalability remains, and the FIBs of core networks may not be aggregated.

A revolutionary approach involves changing both the semantics and syntax of the IP address, redesigning the address structure, and establishing a new routing architecture.

The root cause of problems with routing architecture is the dual semantics assumed by the IP address. The IP address represents both the host's identity (used to represent the endpoint of a session at the transport layer) and the host's location (used for routing data packets in the routing system). According to Rekhter's Law, in order to ensure the scalability of the routing system, IP address allocation should adapt to the network topology. However, because the IP address also represents the host's identity at the same time, address allocation is often based on the organizational structure rather than the topological structure. As such, it is relatively stable and cannot be dynamically adjusted according to changes in the network topology. These two roles have conflicting aims, thus making it difficult for a single IP address namespace to meet the requirements of both roles at the same time, thereby resulting in routing scalability issues. Semantic confusion can cause issues of mobility, multi-homing, traffic engineering, and security to become more complex. Therefore, the starting point of the revolutionary approach is to redesign the address structure.

A notable example of this revolutionary approach is hierarchical exchange. At present, the Internet uses a Mesh structure that does not have a center or hierarchy. Disorderly network architecture makes the addressing of IP packets dependent on routing technology. By redesigning the address structure to make the Internet hierarchical, and associating the address with the network structure so that routing is replaced by the exchange of address fields, problems such as routing scalability will naturally disappear.

Another example of the revolutionary approach is to clarify the location information in the address. The address structure is modified to contain information on both the geographical position and potential energy of the node. When seeking a path, the node does not have to depend on information from the burgeoning routing table. Instead, location and potential energy information of the node are used to determine the route, thereby solving the problem of routing scalability.

Although the starting points of the revolutionary and the gradual approach differ, the two are not contradictory. Although the revolutionary approach is not yet mature, it will have a positive impact on the gradual approach. Meanwhile, the gradual approach provides an important point of reference on how to establish a new architecture, improve innovation efficiency, and avoid lengthy detours. It is a unity of opposites. As far as routing scalability is concerned, the core ideas behind both the revolutionary and the gradual approach are to separate the core and the edge, and the identity and the location. They may well converge in this direction.

5.1.5 Establishing a democratic management mechanism for Internet resources, to solve resource-management problems

At present, there are numerous problems with resource management on the Internet. A single address assignment model results in disorder, while an inability to reflect location information leads to semantic ambiguity in IP addresses. Domain name management is controlled by developed countries, with the ICANN management center located in the United States. All of the root servers are located in developed countries, which this is a source of hidden concern for Internet security in developing countries.

The Internet is not just about productivity. In fact, it can be said to be the third currency, alongside money and energy. It has to reflect the mainstream consciousness of the market, embody the core interests of the world, and promote the spirit of human innovation. If hegemony is to be avoided, a democratic management mechanism must be established. Generally, Internet resource management has the following four development models:

1) Continuation of the community governance model helmed by ICANN

The core of this model is to maintain the analytical mode of the root servers led by the United States. Its limitation is restricted expansion into sovereign and privacy markets. If the international community wishes to prioritize the development of the market for the Internet as the first goal, this model needs to advance with the times and deal with various challenges and changes.

2) A model of joint governance by multiple international organizations

This means switching from ICANN's "one country one organization" management model to an international, multinational model that involves multiple organizations. The core of this model is the expansion and automatization of root server management through different international organizations to become compatible with networks that have security considerations such as energy networks, financial networks, and environmental networks. The benefit of this transformation is that the limited democratization of Internet management is traded for the expansion of the market for the Internet.

3) Learning from the ITU's allocation and management procedures for numbering, naming, addressing, and identification of resources to establish a mechanism for the interconnection and interoperability of Internet domain name management based on sovereign countries

For example, reference can be made to the international long-distance telephone area-code allocation model based on the ITU's E.164 standards. The core of this model is to establish a loosely-organized root server management model based on countries or regions. The difficulty is how to launch and manage such a system.

4) Reorganizing the existing Internet to a governance model of management by the major powers

For example, a coordination mechanism based on the major powers or an alliance of the major powers that are the Internet centers of the world; or reorganize into a cooperation mechanism based on the G20 countries; or for a model in which responsibilities are taken by the major powers and based on a combination of other mechanisms. The core of this model is for the existing community governance model (with ICANN as the main operating body and the United States is behind it) to evolve into a system in which the major powers are responsible. The advantages of doing so are the democratization and marketization of the Internet. In particular, a market is implemented for the accumulation of capital for the upgrading of the Internet and the expansion of new technologies.

The four models above reflect the sovereign interests that correspond to sovereign states or international coalitions, and correspond to the basic characteristics of the Internet as a global shared property like the oceans and Outer Space. These four models are more of a development framework model, and the eventual development may be a composite of several. Alternatively, several models can be reorganized and merged into a step-by-step implementation strategy that addresses issues such as a philosophical reflection of what the Internet should be, a scientific judgement of what the Internet will become, and the innovation resources that the Internet can provide.

5.2 Service support capabilities of future networks

In the initial phase of designing the IP, only the issue of point-to-point communication between hosts was considered. Thought was not given to more general communication modes (such as point-to-multipoint or broadcast), more complex network structures (such as multiple access for users), and more diversified service requirements (such as terminal mobility). Therefore, after these diversified needs emerged, numerous solutions for improvement were also mooted. However, these improvisations are only makeshift measures that do not fundamentally solve the problems. Instead, they bring with them a certain degree of complexity. At the same time, from the perspective of a revolutionary approach, although there are many innovative ideas, they cannot be implemented in the short term. This section combines the two approaches to propose some feasible solutions to the problems with network service support capabilities.

5.2.1 New addressing and routing mechanisms to solve the multi-homing problem

Multi-homing means that the user's network is simultaneously connected to two or more operator networks through the user's address out of consideration for security or quality of service. In this case, multiple routes will appear at the backbone layer of the Internet, increasing the pressure for routing scalability there. The statistics show that more than 30,000 companies and organizations worldwide currently use multi-homing access methods. This is a huge burden for Internet routing. At the same time, changes in these client networks result in frequent oscillations to routing at the backbone layer. At present, the global routing table is updated six times per second, or 500,000 times each day.

For the multi-homing problem, some routing-level solutions have been proposed based on current IP routing technologies, such as the automatic route injection method and the tunnel encapsulation method. Although these methods can somewhat reduce the burden on routing at the core layer of the Internet when all users are connected, judgment and processing mechanisms need to be added to the existing routing protocol, increasing the workload of the equipment during processing.

The essence of the multi-homing problem is the issue of routing scalability at the core layer of the Internet. Innovative Internet architectures (such as hierarchically switched networks, Future Packet Based Network or FPBN, and I3) propose to fundamentally solve the routing scalability problems at the core layer of the Internet by changing address semantics and structure (such as hierarchical addresses) and altering routing methods. This also reduces the impact of the multi-homing problem. From this point of view, it seems that the solution to the multi-homing problem ultimately hinges on new addressing and routing mechanisms.

An improvised solution to the multi-homing problem is shown in Figure 5-7.

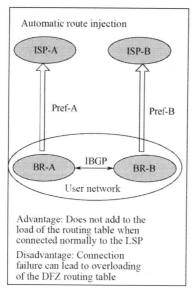
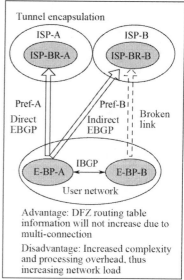

Figure 5-7: An improvised solution to the multi-homing problem

5.2.2 Application-layer multicast to solve the multicast problem

In the initial phase of designing the current Internet, only the issue of point-to-point communication, i.e. unicast, was considered. Point-to-multipoint and broadcast services were not given due consideration. After the emergence of point-to-multipoint and multipoint-to-multipoint services such as IPTV and videoconferencing, a multicast routing protocol based on a unicast routing protocol was designed that used the shortest path tree generated by unicast routing to generate multicast routing, i.e. using unicast to resolve the multicast problem. The disadvantages of this approach are that the multicast aggregation point and the traffic replication point are under pressure, and the multicast protocol itself has limited flexibility. The Protocol-Independent Multicast-Dense Mode (PIM-DM) is based on the method of flooding and pruning, and is less efficient. In the Protocol-Independent Multicast-Sparse Mode (PIM-SM), joining is explicit but there is a greater delay in joining and leaving. Currently, in order to ensure the quality of multicast services, static multicast is used more often, to directly push multicast traffic to all network-edge nodes. However, this method also results in a waste of network bandwidth.

Among the various revolutionary approaches, numerous new network architectures have been proposed, such as NDN and I3. In these networks architectures, the sending and receiving processes are separated. Point-to-multipoint is used as the basic data transmission mode. The sending node only sends data into the network, while the receiving node applies for data and resources according to its needs before the nodes in the network send it the data.

The revolutionary approach solves the point-to-multipoint problem by an innovation of the routing mechanism. However, point-to-point communication is now the more difficult problem to solve. In this regard, there is still no proper solution based on separating the sending and receiving processes. Meanwhile, some solutions have been proposed to solve the problem of point-to-multipoint communication at the application layer, such as CDN and P2P. The emergence of these application layer techniques provides us with another possible approach, which may become the focus of research in future.

Figure 5-8 shows the data transmission modes of the NDN and I3.

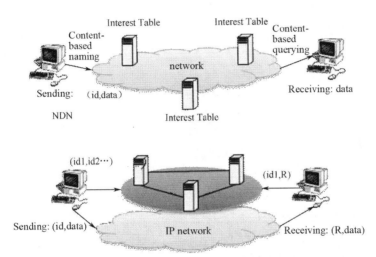

Figure 5-8: Data transmission modes of the NDN and I3.

5.2.3 Application-layer implementation to solve the problem of mobility support

The emergence of wireless data access technologies has allowed the terminal to access the Internet freely without being connected to a cable. This has spawned new terminal mobility needs, as users expect that their applications (especially critical ones such as financial transactions) will not be interrupted when the terminals are on the move. At present, the IP address is both the address of the node and the name of the application. As such, the most basic requirement of separating names from addresses for node mobility cannot be met. In addition, the DNS mechanism (the name-address mapping mechanism in use) is too slow (taking 8–12 hours to be refreshed) to satisfy the high-speed address-name mapping required for rapidly moving nodes.

To solve the mobility problem, Mobile IP appeared several years ago. In this solution, triangular routing is used to allow the data packet to pass through the host's "hometown site" before passing through the tunnel and being forwarded to the roaming area. This solution is inefficient and the process complex. As such, it is not widely used.

Among the various solutions in the revolutionary approach, one basic idea is the separation of applications or resource names from the node address. Meanwhile, efficient mapping techniques (such as DHT) are used to rapidly update the mapping between names and addresses, thereby satisfying node mobility needs.

It should be noted that separating names from addresses is the key to solving the mobility problem. However, in terms of actual requirements, there are few applications that impose stringent requirements on node mobility. Most applications that are insensitive to quality or the connection process can actually provide stable service for the user when the node moves to a new network by maintaining the state of the client and initiating retransmission.

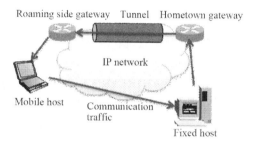

Figure 5-9: The mobile IP protocol process

5.3 External capabilities of future networks

5.3.1 Security and credibility

In its early days, the Internet was thought to be a secure network because most of its default users were communities with common goals and mutual trust. As such, security issues were not taken seriously by its designers. However, with the commercialization of the Internet and its explosive growth worldwide, its user base now differs in terms of scale, goals, and quality. Mutual trust among users has also ceased to exist. Viruses abound, and cyber-attacks occur frequently. Privacy cannot be guaranteed. Now, the Internet is deemed an unsafe network.

At present, although academics and industry practitioners have proposed some solutions (such as DNSec to solve DNS security problems, IPSec to solve end-to-end communication confidentiality problems, SAVA to provide real source addresses, and DFI/DPI techniques for service analysis and identification), because the original Internet architecture lacks an overall security framework, the addition of various security measures in a patchwork manner means that security vulnerabilities or functional overlaps are inevitable.

The most basic requirement of the various solutions under the revolutionary approach is to guarantee the security and credibility of the network. At the same time, they also propose special architectures to ensure the security of the network. However, based on current research, these "security architectures" are still unclear, and difficult to implement in the short term.

Therefore, to cope with the existing network problems, there is a need to learn from the revolutionary approach to design workable solutions based on enhancing network control capabilities. Some basic ideas are as follows:

1) Building a top-down architecture for network security

The functions of terminals, networks, and applications should be clearly distinguished. Functional entities with different functions should then be provided with different levels of security assurance and solutions.

2) Providing security through the isolation of resources

Instead of striving for security in all processes throughout the network, resource isolation is used to segregate services with different security needs so as to provide security on different planes. This satisfies services needs as far as possible without having to deal with the differing impacts on various services.

3) Security techniques targeted at specific services

This involves prioritizing research into security techniques targeted at the specific needs of certain critical services, such as the Internet of Things and cloud computing.

4) Reducing end-to-end IP transparency

End-to-end IP transparency is an important cause of Internet security problems. The use of technical measures to reduce network transparency allows the network to sense and control user services to a certain extent. This helps to improve network security. Nonetheless, network neutrality and fairness are also issues that need to be considered.

5.3.2 Assuring service quality

In the traditional Internet architecture, the core networks only store limited management information and do not provide QoS guarantees. With the commercialization of the Internet, a large number of real-time applications have emerged. Multimedia applications such as video conferencing and IPTV have strict requirements in terms of delay, delay-jitter, and bandwidth. Correspondingly, the network is required to provide an assurance of service quality. The existing QoS guarantees models such as IntServ/DiffServ, Multiprotocol Label Switching (MPLS), and QoS routing, and can be said to be "Band-Aids" for the network architecture. As such, there are problems in their practical application and deployment. For example, IntServ has poor scalability and high management overheads, while DiffServ is unable to provide QoS guarantee for every data stream.

1) The "best effort" mode is unable to support the long-term sustainable development of the Internet

Service quality is a long-term problem that plagues the development of the Internet. After more than 20 years of research, many techniques, standards, and solutions for service quality have been proposed, but on the whole, the "best effort" approach is still used. At present, the Internet is more than capable of supporting the development of video services. This is due to the rapid growth in capabilities and levels of network infrastructure in recent years, as well as improved capabilities of network processor chips and network equipment, and constantly improving network management. However, it should be noted that at this point in time, the overall development of the Internet is still in an early stage. As penetration continues to rise and the Internet becomes more integrated with production and everyday life, services will change rapidly, and traffic will grow quickly.

Soon, Internet traffic will grow at a faster rate than that of its network infrastructure capacity. As a result, traffic will be in short supply, and using the "best effort" approach to manage it will not be sustainable.

Firstly, the "best effort" mode does not conserve energy and is not environmentally friendly, since it simplifies the protocol and implementation, and results in uneven resource utilization and wastage. Second, this mode is not always effective because statistical multiplexing will inevitably lead to local bottlenecks, and local congestions will render it ineffective. Finally, the "best effort" mode is unable to satisfy integrated service needs since it does not support service differentiation.

2) The intrinsic nature of Internet service quality problems

The root cause of failure after more than 20 years of hard work is that Internet service quality problems are intrinsic and closely related to core technologies such as addressing and routing. To fundamentally resolve problems with service quality, it is necessary to modify the Internet's "genetic make-up", i.e. its addressing and routing mechanisms.

(1) The statistical multiplexing feature of the packet-switching network— Link utilization is severely imbalanced, and it is difficult to implement bandwidth design and precise bandwidth-control for links.

(2) The uncertainty of communication paths—Disordered network topology and dynamic non-deterministic routing. For the connectionless mode, the uncertainty of paths makes it impossible to perform resource reservation according to the path. For example, the DiffServ technique can only achieve node-level resource management, and lacks the necessary admission control. Although QoS Routing and Backup-Path Routing solve the issue of path QoS, the associated costs are high. For the connection-oriented mode, the cost of resource reservation and management during connection-establishment is too high. For example, RSVP and MPLS-TE were not successful because of their complexity.

(3) The uncertainty of service traffic—The two major conclusions of "power-law" and "self-similarity" in the study of Internet network behavior have no practical significance. Spikes in traffic are an inherent characteristic of packet-switched networks. Networks and services are unable to sense them. The complete separation of services and bearers has resulted in innovations in Internet services, but has also restricted the implementation of QoS.

3) The key to solving Internet service quality problems is to strike a balance between flexibility and manageability

The solution to future Internet service quality problems lies in balancing the flexibility of statistical multiplexing with the rigor of meticulous control. On one hand, we want to preserve the flexibility of service brought about by the "best effort" approach, and

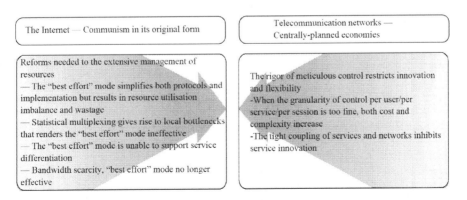

Figure 5-10: Tackling QoS issues requires a balance between flexibility and manageability

the separation of networks and services. On the other hand, we must consider the high efficiency brought about by refined resource management.

Tackling QoS issues requires a balance between flexibility and manageability, as shown in Figure 5-10.

4) Architectural innovation is needed to solve future Internet service quality problems

Since Internet service quality problems are due to the genetic make-up of the Internet, architectural innovation is needed to ensure that Internet resources are knowable, manageable, and controllable. This can be achieved by investigating new types of addressing and routing mechanisms.

Structured addresses and hierarchical network structures are feasible ways to solve future Internet service quality problems. There is no longer a need for global routing since routes can now be confirmed. The logical layering of the network also means that the resources at each layer are relatively independent. This in turn allows the implementation of layer-based admission control and traffic scheduling to effectively manage the level of complexity.

5.3.3 Energy conservation

As the scale of the Internet continues to expand, elements including network equipment, terminal equipment, and IT systems consume a growing amount of energy. The statistics indicate that energy consumption by the Internet was 868 kWh in 2007, accounting for 5.4% of the global total. Internet energy consumption in the US already accounts for 9.3% of its total national energy consumption, and is growing at an annual rate of 8% to 10%.

In the face of increasing pressure to conserve energy, there is a need to look into energy conservation and consumption reduction for the Internet from various aspects. These include terminals and equipment, network technologies, service platforms, and network layout.

For terminals and network equipment, energy-saving processor chips can be used, and system software with energy-saving features can reduce consumption. At the network layer, reducing the circuitous paths taken by traffic and removing redundant or dormant paths can effectively reduce energy consumption during data transmission. From the perspective of services, the use of cloud computing, virtualization, and other technologies can improve the efficiency of resource utilization and slow the growth of energy consumption. In terms of network layouts, it is necessary to coordinate with energy resources so that network infrastructures with high energy consumption levels (such as data centers) are clustered in resource-rich areas to improve resource utilization efficiency. This way, "bits" will be transmitted instead of "Watts".

Various aspects of energy conservation and consumption reduction for the Internet are shown in Figure 5-11.

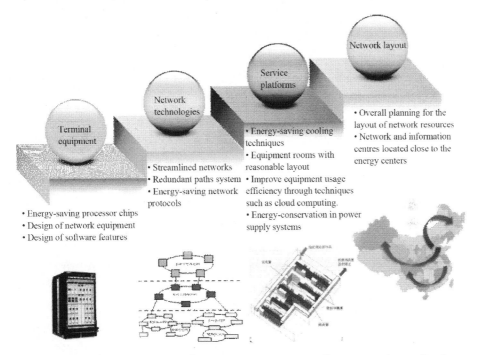

Figure 5-11: Various aspects of energy conservation and consumption reduction for the Internet

5.4 Research into new network architectures - current status and trends

Based on the analysis in the previous sections, it is very difficult to resolve the majority of the Internet's problems through improvements or patches. Therefore, China must bolster its research into innovative network architectures. Technological breakthroughs should be sought, with addressing, naming, and routing as the main thrusts and based on key elements in future network technologies.

When it comes to Internet architecture, there are numerous schools of thought both domestically and abroad, and the technology is developing in more than one direction. As such, China should maintain its technical diversity and concentrate on supporting research into three to five types of new network architecture and the relevant protocols. At the same time, technical solutions in China that have been developed to a suitable extent should also be selected to have prototypes developed for experimental verification. Lastly, the coexistence, competition, and development of the various architectures should also be promoted.

Research into future Internet architecture should reflect the openness of both the protocol and the architecture, with attention paid to the standardization of the network-operation interface that will facilitate the flexible design, development, and deployment of new network protocols. Meanwhile, from the perspectives of manageability and operability, the security requirements of authentication, auditing, traceability, filtering, and intelligent prevention should be incorporated into the top-down security design of the next-generation Internet.

Finally, China should also pay more attention to promoting its independently-developed core technologies abroad. It should participate in the formulation of international standards for the next-generation Internet by pushing its own domestic standards, with intellectual property rights to be adopted as the international standards.

5.4.1 Existing problems with the network architecture

With increasing demand for information services, the information-driven society of the future will expect credible, secure, and reliable network and communication services. However, in the face of such demands, many problems and deep-seated conflicts with existing networks have been exposed. This shows that existing networks are unable to match the sustainable development of the future information-driven society.

1) Problems with the Internet

The Internet cannot guarantee basic security and credibility. It is uncontrollable and unmanageable, and cannot provide QoS guarantees.

(1) The untrustworthiness of the Internet is manifested in all aspects of its design, construction, and operational management. The frequent occurrence of security incidents is a concrete manifestation of the its vulnerabilities. These security issues have greatly limited the possibility of in-depth application and service innovation, and have also severely restricted the Internet's development and the utilization of its enormous potential. At the same time, it has also affected the healthy development of national economies, and has even threatened social stability and national security.

(2) The Internet is essentially a "best effort" connectionless service without any QoS assurance. In a data service environment that is predominantly about FTP, E-mail, and Web services, the Internet is able to meet the basic needs of its users. However, in the face of real-time delivery of streaming media such as voice calls and IPTV, and

the emergence of a large number of mobile communication terminals, the existing Internet is unable to provide adequate performance, functions, and the necessary QoS.

One of the core ideas of the Internet is the memoryless transmission of data so that as little information as possible is stored in the network, thereby ensuring the simplicity and efficiency of network devices. This means that network elements lack the information needed for management, making it impossible for administrators to manage and control the network efficiently.

2) Problems with telecommunication networks

Traditional telecommunication networks are inadequate when it comes to integrated multi-service bearing and the flexible deployment of new services. The development of the next-generation network (NGN) with IP as the core is plagued by the intrinsic deficiencies of the Internet.

Existing telecommunication networks are implementing the technological transition of bearer networks by making use of IP as their core. Undoubtedly, service networks will be based on bearer networks that use IP as the technological foundation. However, the existing IP bearer networks cannot be controlled and managed. They are unable to provide QoS guarantees, basic security, and credibility assurances. What's more, encryption technology is misused and out of control, posing a serious threat to the secure communication of information. In addition, the design of IP bearer networks means that they are unable to adapt to new service-bearing needs arising from changes in the network's traffic model.

At present, telecommunication services are based on unreliable IP networks, and a large number of unreasonable functions exist in such service networks. There is also a lack of innovative services with Chinese characteristics. When it comes to thinking about service implementation, the telecommunication industry differs from the Internet industry. The approach of each has its strengths and weaknesses. At present, there is a lack of integration and coordination. At the same time, how to go about implementing new services in future is also unclear, and there is much room for improvement in terms of the maturity and reliability of the service networks. Service implementation methods are diverse, chaotic, and volatile. As such, there are difficulties in interconnectivity and interoperability; multiple service networks built separately are unable to offer flexible service combinations, thus impeding the provision of personalized and integrated services to users. Service networks are susceptible to attacks, and have inadequate traceability capabilities, leading to security risks. Compatibility between the service networks and bearer networks is relatively weak, system efficiency is low, and flexibility is poor. The segregation or association of user data is complicated; SP/CP access has not been standardized for multi-service platforms, and there is no open and convenient service development environment, making it difficult for the logic and content resources of third-party applications to be reused. The capability of value-added service platforms is relatively static, and there is no unified service management platform to integrate and manage multiple service capabilities. In addition, the development of the operation and maintenance system of existing networks lags behind that of both bearer and service networks.

3) Problems with cable networks

The cable television network is an important part of the national information infrastructure. However, its current service form of one-way-pushing and a single profit model can no longer satisfy consumers' needs for multiple forms of entertainment. At present, it faces the daunting tasks of digitization, transforming to become bidirectional, expanding its information service capabilities, and updating its business model.

Regarding digitization, after settling the argument of whether to transform or not, the issue now is how to go about the transformation. Digitization of the traditional analogue television network is already underway in China. During a process of gradual exploration, various development models such as the "Qingdao Model", "Hangzhou Model", and "Shenzhen Model" have been created. Following pilot projects in some cities, the digitization of cable television services is presently being rolled out to other large and medium-sized cities across China. At the same time, networks are remade with larger capacity, to be both bidirectional and multi-functional. Network services are being expanded, and the construction of a full-featured digital cable television network is being promoted.

5.4.2 Status and trends in international research

Network architecture is a general concept. Researchers with varying backgrounds and research purposes have different understandings of its boundaries. Nonetheless, the common consensus is that network architecture mainly includes a functional model, a topological model, and an application model, of which the functional model is the core. The functional model defines how the network system is partitioned into smaller parts with different functions, and how these parts interact, as well as how the permutations and combinations of these parts are used to achieve the overall functionality of the network system.

At present, global research in this field can be divided into two parts. In the first, the basic scientific questions about the new types of network are solved from the perspective of fundamental theoretical network research. In the second, specific problems in network and service implementation are resolved from the perspective of technical and engineering research. These two parts affect each other and develop interactively.

At present, there are two important directions in the fundamental theoretical research of new network architectures: the improvement of the existing hierarchical network functional model, and the exploration of non-hierarchical network architecture.

1) Improving the hierarchical functional model

There are two important trends in the improvement of the existing hierarchical functional model. One is the simplification of the hierarchical model, i.e. flattening the network hierarchy, while the other is cross-layer access and multilayer iteration in the hierarchical model.

(1) Flattening the network hierarchy

The proposal for the hierarchical functional model for the telecommunication network is a technological breakthrough in the field. The ISO's OSI seven-layer protocol stack model, the Internet's TCP/IP four-layer protocol stack model, and the ITU-T's X.200 hierarchical model (the X.200 adopts the OSI model and re-standardizes its concept according to the ITU-T perspective, and has become a widely-used network model and network-design method) break the convention of not separating the networks and services in telecommunication networks. The abovementioned models have also separated service implementation from specific network technologies, boosted integrated services, directly brought about the rapid development of telecommunications networks and the Internet, and promoted the informationization of society.

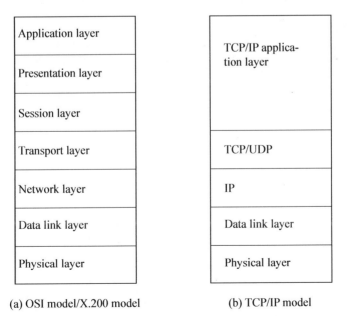

(a) OSI model/X.200 model   (b) TCP/IP model

Figure 5-12: OSI model/X.200 model and TCP/IP model

The OSI model/X.200 model and the TCP/IP model are shown in Figure 5-12.

However, with the expansion of the network and the growth in user numbers, as well as the constant demand for new services (especially the establishment of the direction in which tri-network integration would develop), more of the defects and deficiencies of traditional hierarchical functions are being exposed.

First, excessive network layers mean that complex functional interfaces need to be defined thus increasing the complexity of implementing network protocols and systems, reducing the efficiency of the system. Second, there are problems such as functional overlaps between the layers, unclear demarcation of functions, and expansion difficulties due to fixed layers. Reflection on these issues by the industry has turned the flattening of the network hierarchy into an important research topic. However, results differ on how to simplify the network hierarchy.

Currently, the information communication architecture (ICA) as defined by the ITU-T Y.130 represents the structural evolution of an iconic networking technology in the field of information communication. It proposes a three-layer structure for future networks, namely the application layer, middleware layer, and baseware layer. Both the middleware and baseware layers are considered part of the network infrastructure.

In the ITU-T's NGN model, the network hierarchy has been simplified to only two layers: the service layer (mainly based on IMS) and the transport layer (research into the future packet-based network [FPBN] is currently underway). The FARA (Forwarding Directive, Association, and Rendezvous Architecture) model proposed by the Massachusetts Institute of Technology also simplifies the network hierarchy and proposes a two-layer network structure. Streaming media broadband network (Medianet Protocol, or MP), the next-generation network platform that is positioned to provide multimedia services, also divides the network into two layers: the network layer and the media layer.

Typical structural models are shown in Figure 5-13.

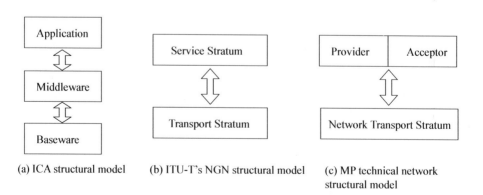

(a) ICA structural model   (b) ITU-T's NGN structural model   (c) MP technical network structural model

Figure 5-13: Typical structural models

(2) Iterative network architecture

With the increase in the number of network technologies and the acceleration of upgrades to network equipment, several iterations of network architectures have emerged in actual network construction, such as IP over MPLS over IP over ATM over SDH over DWM. The iterative architectural relationship that was originally quite clear has now become complex. Multi-layer iteration requires that each layer has a unified functional interface with all other layers. This is different from the original ISO OSI seven-layer model in which the lower layer only has to provide an interface for its adjacent upper layers. Therefore, a hierarchical network model supporting cross-layer access has emerged.

The main objective of the cross-layer model is to integrate the various network layers through defining a flexible and scalable network architecture in order to establish a new model that provides flexible network resource sensing, analysis, decision-making, and scheduling. Doing so would change the conventional, hierarchical network model in which each network layer is only able to exchange information with layers adjacent to it, in which there is a lack of mechanisms for correlation analysis and comprehensive decision-making, and in which there is a lack of comprehensive utilization of the resources belonging to each layer. Therefore, the network has greater flexibility and scalability

However, this cross-layer model puts forward higher requirements for the design of each network layer. The format of both the input and output information and calling functions at each network layer need to be standardized. In addition, cross-layer direct access and call at each network layer can only be achieved when the external interface of each network layer is standardized. These standardized external interfaces actually form a single logical bus instead of two buses, i.e. the input bus and the output bus

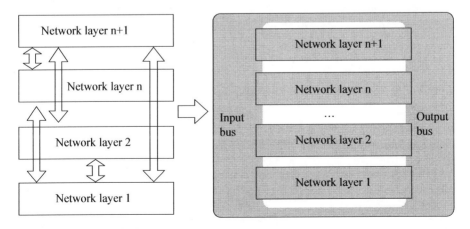

Figure 5-14: The cross-layer access model

in the functional model. Each network layer is plugged into this logical bus, so each can call functional modules from other layers through the bus, thereby achieving the flexible calling and combining of multiple network layers.

The cross-layer access model is shown in Figure 5-14.

In fact, the cross-layer direct access model is a departure from the conventional hierarchical model in terms of the layer adjacency relationship, meaning that each network layer now has greater autonomy and can support more flexible network-layer iterations. For this reason, ITU-T has proposed a set of new methods for the design and analysis of new types of network, as in ITU-T G.805 and G.809. In G.805 and G.809, the network-layer communication mode is abstracted as connection-oriented and connectionless, and the iterative relationship between different network layers is transformed into a coupled iterative relationship that falls between Connection-Oriented Packet-Switching (CO-PS) and Connectionless Packet-Switching (CL-PS).

2) Non-hierarchical network architecture

Both the ITU-T G.805 and G.809 actually reflect a new way of thinking in network design. Through the refinement and flexible combination of functional modules, a more scalable network architecture is achieved. However, the functional modules refined by G.805 and G.809 retain a hierarchical structure as they are geared towards CO-PS and CL-PS. With the in-depth development of this research direction, methods to refine and combine other functional modules will naturally emerge, and some new non-hierarchical network architectures will appear. These include theoretical models such as object-oriented network architecture, role-based network architecture, and service-unit based network architecture.

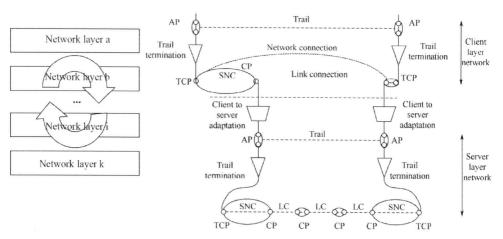

Figure 5-15: The multi-layer iterative model

The multi-layer iterative model is shown in Figure 5-15.

3) Network architecture

(1) Object-oriented network architecture

This new network architecture adopts object-oriented technology from the computer field. Through functional aggregation and abstraction, a number of functional classes are defined to form a toolbox. Tools (i.e. functional network modules) in the toolbox can be flexibly combined, with the outcome that network functions are now more flexible. These classes can give rise to a variety of tools that meet specific needs. The instantiation of classes can meet personalized needs.

The object-oriented network architecture is shown in Figure 5-16.

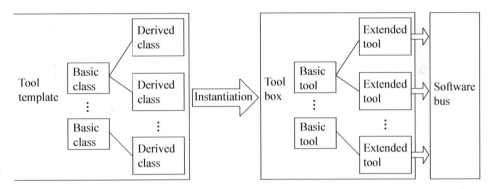

Figure 5-16: The object-oriented network architecture

At present, this research direction is still in its infancy. Focuses include toolbox definitions (abstraction and encapsulation of network functions) and the combined interaction mode of the tools (bus structure and call mode). The different ways of combining tools and collaborative working methods correspond to different new network types. At present, only one network design and implementation idea has been proposed, and there has been no actual implementation.

(2) Service unit-based network architecture

In response to inefficient network services caused by the overlapping functions of different layers, and the complex layered-processing of the existing hierarchical network architecture, this model changes the relationship between the serving layers and the served layers in the hierarchy model. Network service functions are split into service units that form a service team, which provides services to the application groups. The units only provide and do not accept service, thus avoiding the overhead of inter-layer interaction and service delivery. A service unit can do more than just serve the applications of a node. Service units from different nodes can cooperate to serve a particular node or the entire network. In fact, each service unit is an independent functional module with a standardized interface for input and output, and is able to complete a piece or multiple pieces of independent work. The service unit is the smallest entity that can provide network services while keeping internal details hidden.

A network established based on this network architecture has an end-to-end virtual circuit structure. Depending on the type of service (real-time audio or video, or text data and priority level) needed, the same source or destination node address uses different service units. Different virtual circuits are then built for transmission, to ensure service quality and security. China has supported this research direction in the 973 program.

(3) Role-based network architecture
In their 2002 article entitled *From Protocol Stack to Protocol Heap-role-based Architecture*, Braden et al. proposed a role-based network architecture design that does away with layers. By clarifying the entity (subject) and its role in the information communication process, the communication process is transformed into a collaborative work process involving subjects with different roles. The role-based network architecture analyzes role-oriented work models and design methods from the perspective of roles and cooperation.

The role-based network architecture is a generic architecture that includes a variety of network architectures. An example would be the I3 (Internet Indirection Infrastructure) – a new type of Internet architecture that emerged in the last two years, as well as its extension (the Secure-I3) and the Hi3, which incorporates HIP technology. I3 is an application-layer, cross-layer network used for forwarding packets. It is based on P2P technology and was proposed by Ion Stoica et al. from the University of California, Berkeley. The I3 is an architecture that departs from the end-to-end communication model of the existing Internet because it introduces "indirect" into the communication mechanism. Through the "indirect" mechanism, the core network functions are brought into full play. There are three main roles in I3: senders of information, receivers of information, and triggers of information. At present, the sending and receiving of network packets is a standardized process. Packets sent by the source end will normally be received by the destination end. However, sending and receiving are independent from each other in I3. The receiving party informs I3 of its location and the data characteristics that the required packet should have before I3 forwards a packet with such characteristics to it. Similarly, the sender does not need to specify the recipient of the packet, but only needs to label the data characteristics of the packet before handing it over to I3. Since the sender does not need to know the identity and location of the receiver at all, so long as the mobile node provides I3 with timely updates of its location, it will receive the packet correctly. For a group of users, each will receive an identical packet if it conveys the same data characteristics to I3. Therefore, I3 is able to support host mobility and multicast. For example, if the receiver (host R) inserts a trigger (id, R) into the I3 architecture, then the host R can receive all data packets with the identifier "id". Role-based network architecture is shown in Figure 5-17.

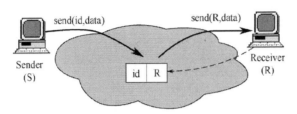

Figure 5-17: Role-based network architecture

Due to the appearance of the trigger, the restricted route can be implemented. For example, in the trigger (id, R), "id" can be a sequence such as {id1, id2, id3}. In this case, the sequential transfer of the data packet between "id1", "id2", and "id3" is thus supported. I3 can also provide architectural support for host mobility.

In addition, the Medianet Protocol (MP) technology proposed by overseas Chinese technicians in the United States can also be considered as a new type of role-based network architecture. It targets video communication services and divides network roles into "receivers", "transmitters", and "storage". These roles are then used to design both the functional and the networking models of the network to support broadcast, multicast, and on-demand and media-program scheduling functions, as shown in Figure 5-18.

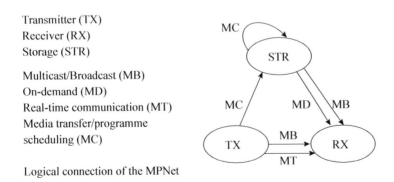

Figure 5-18: MP technology

4) Trends

Based on the above analysis, current research on new network architecture is heading in the direction of flatter structures, multi-layer iterations, and flexible function deployment. These developments are not isolated or mutually exclusive, but are gradually being integrated.

In future research, the hierarchical model will still have instructional value, but non-hierarchical network architectures that are role-based or object-oriented will have an increasingly important role to play. There is still room to integrate the research directions in these few network architectures.

When designing a new type of network architecture, it is not necessary to follow the traditional hierarchical design method. Instead, starting from the application scenario, the roles and the collaborative relationships between them are clarified to construct a role-based model in the bearer network and service network respectively. On this basis, the functions necessary for these roles to collaborate are determined in order to build a "toolbox". Different combinations of the functional modules in these toolboxes correspond to different network architectures. In fact, it is even possible to further explore the bus architecture (or other possible tool-assembly methods) in which these tools work together to implement a common design platform for new network architectures, on which new network types can be designed more flexibly.

For example, in the design of a new bearer-type network, it is not necessary to separate the network layer from the link layer at the beginning before designing them separately. Instead, after determining the roles and the communication requirements in the bearer network, planning is carried out to decide the functional modules to be included in the new type of bearer network, as well as whether each module is basic or derived. Then, the external interface of each module is standardized. There are many ways to implement a functional module, such as the object-oriented method, but this is an engineering implementation technique, i.e. a specific network-function design method. Based on this, the workflow and method to combine a variety of functional modules can be designed to achieve a more flexible and open network architecture. Separate functional plug-ins can be combined into tightly-coupled, large-particled, functional entities known as "functional pieces".

This approach can also be used in the design of service networks for tri-network integration. First, the communication roles of the various services are defined (in research into new service systems, there is a need to focus on video services since they will be the primary means of communication in future) before a process of refining, abstracting, and aggregating is carried out to produce a set of functions. What these functions have in common will form the basic tools and expansion tools for specific services that can be derived from them. In turn, new types of system for specific services can be realized through different combinations of these functional modules.

Therefore, role-based, object-oriented, and other technologies can be used as an effective means of designing new types of network architecture that can be combined with the hierarchical network architecture. In the hierarchical model, after full consideration is given to the problems (such as resource-sensing difficulties, weak cross-layer resource management, poor network manageability and controllability, inefficient matching between network and service, and unreasonable business models) caused by excessive separation of networks and services, a new network hierarchy with the addition of a network layer and a sensing and adhesion layer may appear (see Figure 5-19). Within the trend of separating networks and services, this new network hierarchy can better match services to networks while improving resource utilization and overall efficiency. At the same time, it is also responsible for the construction of a

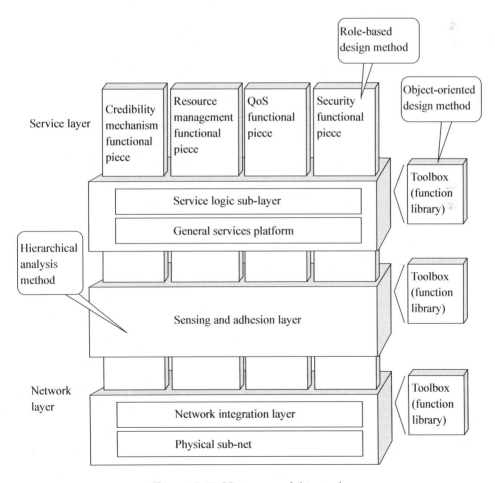

Figure 5-19: New network hierarchy

unified network boundary, to implement the UNI/NNI/SNI interface, and to provide unified access and identity transformation (mapping) to a variety of networks with different structures.

Therefore, in the research into future network architectures, the following areas may become new priorities:

- Research into a model that involves network and service sensing and adhesion, and specific techniques. This is a priority for innovation;
- Design of the "toolbox", i.e. the abstraction and encapsulation of both network and service functions. This is a design priority. The object-oriented technique may have an important role to play;
- The assembly of functional pieces, i.e. the workflow and methods of combining network and service functions. The role-based method is an important way of designing functional pieces.

5.4.3 Status of research in China

China attaches great importance to research and experimentation in future networks. During the 12th Five-Year Plan period, the 863 program planned the Future Internet Architecture Innovation Environment (FINE) and the Software-Defined Network (SDN) Key Technologies R&D and Demonstration projects. In 2012, the NDRC established the CNGI New Network Architecture Technological R&D and Experimentation Project, which includes new address-driven network architecture, technological R&D and experimentation, and evolvable next-generation, high-intelligence network architecture research and experimentation. The 973 program has also supported a number of basic research projects on future networks.

1) When it comes to research into future networks, China has been very successful, with lots of positive outcomes in the areas of basic theoretical innovation, international standards, and industrial demonstrations and applications. It has proposed a number of new network architecture solutions, and has formed a number of outstanding scientific research teams.

The Packet Telecommunication Data Network (PTDN) proposed by the Academy of Telecommunication Research at MIIT (Ministry of Industry and Information Technology) retains all the technical advantages of the existing IP network, addresses future network needs, and solves existing IP network problems. It is a future network solution for which China possesses key technologies. At present, the PTDN has played a major part in formulating a series of ITU standards. A large-scale experimental network platform that is leading the pack has also been built in China, and commercial PDTN equipment has been developed, laying the foundation for the development of the PTDN industrial chain. At present, pilot testing of the PTDN is being conducted in the networks of the Chinese military and PetroChina.

Beijing Jiaotong University has proposed a trustworthy integrated network and universal service architecture, and has defined a two-layer architectural model that includes a trustworthy integrated network and universal service. The network system prototype completed its technological outcome evaluation in December 2009. Following this, the relevant patents have been transferred to the ZTE Corporation, which is integrating the technological solution to develop its own network solution.

The Chinese Academy of Sciences has proposed a hierarchical switching network with the aim of building a high-performance, manageable, controllable next-generation Internet. At present, this technical solution focuses more on theoretical research and verification, and has not been implemented yet.

In addition, the reconfigurable network put forward by the PLA Information Engineering University and the SOFIA project at the Chinese Academy of Sciences carry out research and exploration into future network from the aspects of new network architecture, routing, addressing, security, and network virtualization. The research outcomes produced has improved the mobility, security, manageability, and controllability of the Internet.

2) The main technical direction of China's future network research is basically the same as the mainstream internationally, but the depth is inadequate across the board, and technologies developed domestically are few and far between

Currently, China's research into future networks covers major technical directions such as the identifier system (the separation of identifiers and addresses, and the separation of the periphery and the core), network reconfiguration, separation of control and forwarding, integrated storage for network forwarding, and computing capabilities. However, it is mainly a follower, with little influence and few innovations. Even though China covers nearly all of the research directions, it started late in the development of an experimental environment for technological innovation, and is now lagging behind. The country is yet to build a large-scale future network testbed similar to the United States' GENI or the European Union's FIRE. This is detrimental to openness, collaboration, and innovation in the field of scientific research, and is not conducive to the verification and industrial transformation of research findings.

From the perspective of future trends in Internet research around the world, the network innovation environment (or testbed) is an important means of promoting technological research. China should also pay attention to building an environment for future network innovation by supporting dedicated testbeds for core architecture innovation. At the same time, the country needs to integrate existing testbeds and establish a sustainable and evolving experimental environment through federated means. China should also take a longer-term strategic view when setting up testbeds by looking at the possibility of the test environment becoming the first tier of the next-generation Internet, and exerting influence internationally by being interoperable with existing foreign platforms.

5.5 Test Platforms for Future Networks

5.5.1 The development of future networks has become a strategic orientation for the United States and developed countries in Europe.

Future Internet technologies are still in the early stages of development. There is an endless stream of technologies with a full gamut of features, signaling that this is the peak period for technological innovation. This will directly affect the technological direction of the global Internet in the coming decades. What's more, due to the importance of the Internet as a global information infrastructure, it will have a long-term and profound impact on the political, economic, cultural, production-related, and lifestyle aspects of the future information-driven society.

To maintain and encourage this period of active technological innovation, certain mechanisms and environments should be established. Both the United States and the European Union have incorporated future Internet innovation into their overall strategies in the ICT (information and communication technology) field, and attach great importance to the construction of an experimental environment for innovation.

1) Strategic considerations of the United States

The United States, with its first-mover advantage in the Internet field, has been a global leader in information and communications since the 1980s. During this crucial period of global ICT transformation and industrial integration, especially at the critical juncture of the global financial crisis, the United States is looking to promote both technological and service innovation in information and communication, and seeks to build up a new national IT infrastructure through the construction of an experimental environment for future Internet innovation. By doing this, the US hopes that it will be able to continue charting the global direction of ICT technological and industrial development, thereby maintaining its strategic advantage in terms of technological and industrial strengths, and also its global leadership in ICT. The development of the next-generation Internet is an important part of the country's overall strategy for ICT development.

2) Strategic considerations of the EU

Against the backdrop of accelerating broadband development worldwide, the EU is looking to build on its strengths in wireless and mobile communications in order to combine the next phase of broadband development with the mobile Internet. In so doing, the EU will gradually move away from following in the US's footsteps when it comes to Internet development, and will eventually surpass the US to become the global leader in the Internet field.

5.5.2 Building an experimental environment for future network innovation is an important initiative in Europe and America

Under their respective overall strategic orientations for future Internet development, both the US and the EU regard the establishment of an experimental environment for innovation as an important measure for implementing future Internet development strategies. Each has also built an experimental platform for future Internet innovation with the following considerations:

- To provide a large-scale and comprehensive experimental environment for various innovative techniques through an experimental platform for future Internet innovation, so that innovative network architectures and key techniques can be fully tested and validated, thereby maintaining the vitality of technical and service-related innovation. This would nurture leading technical lights, giving rise to and integrating core technical solutions for future networks so that the experimental platform becomes an incubator for future network techniques.
- Both the European and American experimental platforms for future network innovation are open to anyone, supporting and encouraging the participation of research organizations and individuals from other countries. Thereby, innovative ideas and techniques from around the world will be attracted to the platform, making it a magnet for global talent.
- The Internet of the future will not appear out of the blue. The fact that Europe and the United States are placing emphasis on building an experimental platform for future Internet innovation implies that they see it as a prototype of the future Internet. The experimental platforms in both Europe and America can be the starting point for Internet evolution, with the construction of the ARPANET in the US in 1969 as a successful precedent. Therefore, the overall

structure and layout of the experimental platforms in Europe and United States may directly affect the global network architecture of the future Internet. Based on the current development trend, it seems likely that Europe and the United States will maintain their tier 1 advantage in future networks and networking.

5.5.3 The experimental environments in Europe and United States include four major types of testbeds with three levels

Europe and the United States are very much alike when it comes to design principles, construction methods, and development ideas for constructing experimental environments.

1) A basic design principle for establishing an open and coordinated federation of testbeds

In building its future-network experimental environment, FIRE, the EU is guided by the basic design and construction principle of establishing an open and coordinated federation of testbeds.

In the SPIRAL2 phase of its GENI project, the United States proposed a comprehensive experimental environment with an open federated model that can integrate the various testbeds in the SPIRAL1 phase, conveying its intention to build an open and coordinated federation of testbeds. Strong attention has been paid to this intention, and more concrete implementation is being carried out in the SPIRAL3 phase of the GENI project.

2) Four types of testbeds have been built by Europe and the United States for different experimental purposes

The future network testbeds constructed by the US and the EU are more than just testbeds per se. Instead, they represent the construction of numerous testbeds that can be classified into four categories:

(1) Dedicated testbeds for specific applications
For example, GENICloud in the United States, which is dedicated to experimenting with technical solutions for cloud computing; VITAL++ in the EU tests IMS+P2P technical solutions; WISEBED in the EU is used for sensor networks testing, and CREW is used for testing spectrum optimization techniques such as cognitive radio.

(2) Testbeds based on programmable techniques
Enterprise GENI in the US primarily makes use of the OpenFlow technique to build

a reconfigurable testbed; the OFELIA project in the EU also uses the OpenFlow technique to build a testbed for innovative technologies in the lower layers.

(3) Testbeds based on techniques that can overlap
For example, PlanetLab and PlanetLab2 in the United States, and OneLab and OneLab2 in the European Union, which both use cross-layer techniques such as P2P to build more universal testbeds that support the simultaneous testing of multiple technical solutions. However, these testbeds are often unable to support the innovative experimentation of underlying techniques and are mainly used for the testing of innovative services and applications.

(4) The federation model
For example, the CMULab in the United States and the PII project in the European Union, which are both investigating how existing testbeds, as well as those under construction, can be integrated into a federation so that both dedicated and general-purpose testbeds can be developed sustainably in the long term.

The federation model is shown in Figure 5-20.

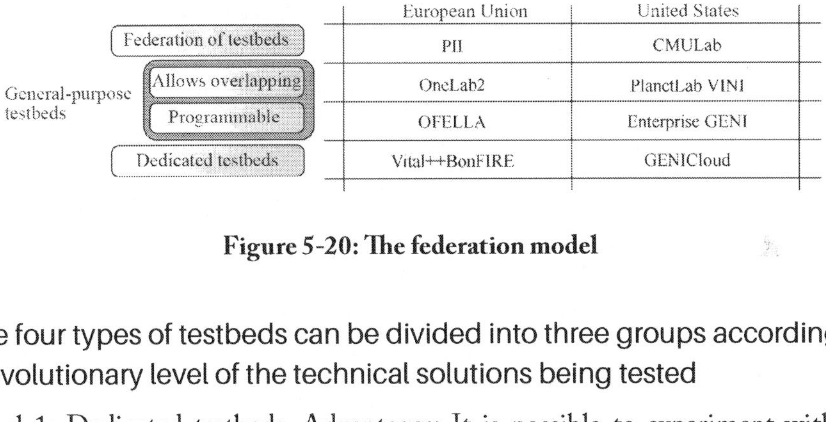

		European Union	United States
	Federation of testbeds	PII	CMULab
General-purpose testbeds	Allows overlapping	OneLab2	PlanctLab VINI
	Programmable	OFELLA	Enterprise GENI
	Dedicated testbeds	Vital++BonFIRE	GENICloud

Figure 5-20: The federation model

3) The four types of testbeds can be divided into three groups according to the revolutionary level of the technical solutions being tested

- Level 1: Dedicated testbeds. Advantages: It is possible to experiment with core technical solutions such as addressing, naming, and routing that usually require specific infrastructure and key mechanisms that are not required by other technical solutions. Disadvantages: They only support the short-term testing of specific technical solutions. They are limited in scale and low in openness.
- Level 2: Universal testbeds. Advantages: It is possible to design a variety of technical solutions in parallel, thus reducing construction costs and usage thresholds. Disadvantages: They are limited to the sort of basic techniques and core infrastructure used in general-purpose beds; only certain L4–L7 technical solutions can be tested, but not those from L2–L3.

- Level 3: Federal testbeds. Advantages: It is possible to integrate and reuse the resources of some of the testbeds that are dedicated to short-term testing, thus allowing multiple independent testbeds to share resources and complement each other. Disadvantages: Similar to those of general-purpose testbeds. They are unable to support the testing of core techniques such as addressing and routing.

5.5.4 The construction of experimental environments in Europe and the United States takes two major trends into account

Analyzing the technical characteristics of the four types of testbeds in Europe and the United States, there are two major trends in the construction of experimental environments for future networks.

1) Innovative architectures require dedicated testing platforms

Underlying core techniques are tested through the establishment of dedicated testing platforms (see Figure 5-21) in order to promote infrastructural innovation.

At present, the EU supported the construction of numerous dedicated testing environments in the FP7. For example, BONFIRE is used to test cloud computing techniques; SMARTSANTANDER is used to test IoT (Internet of Things) techniques in sensor networks; CONNECT is used to experiment with integrated access

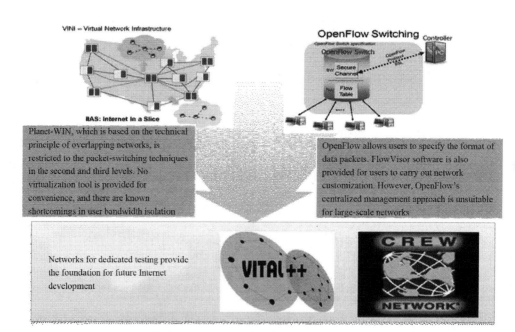

Figure 5-21: Dedicated testing platforms

techniques within multiple wireless techniques; CONVERGENCE is used to test new, content-centric networking techniques based on the subscription/release model; EULER is used to experiment with new distributed dynamic routing mechanisms; HOBNET is used to experiment with new energy-saving networking techniques; and LAWA is used to test large-scale data analysis and data mining techniques.

2) Federations of testbeds are the way forward

Establishing federations for various testbed types (see Figure 5-22) to increase the efficiency of resource usage solves the problem of re-using dedicated networks. The scale of experiments can also be enlarged through collaboration, so that tests which require more than one testbed can be conducted.

There are two ways to go about constructing a federation of testbeds. The first is the central model based on centralization and local de-centralization, while the second is the distributed model based on a federation of equivalent peers.

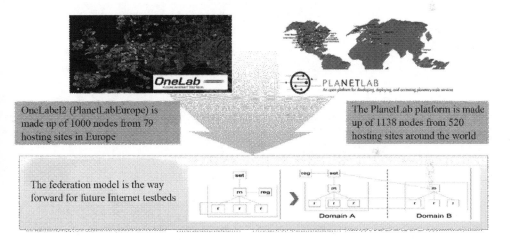

Figure 5-22: Federations of testbeds

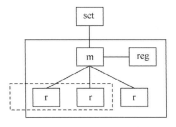

Figure 5-23: Abstraction of resources

Each dedicated testbed and general-purpose testbed can be abstracted into Resource (R), Domain Manager (M), Resource metadata registry management server (Registry, or Reg), and Resource operation set (SET) through the use of a resource integration technique (see Figure 5-23).

The functioning of the central model is shown in Figure 5-24.

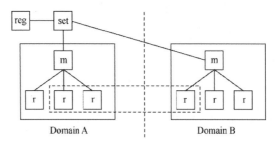

Figure 5-24: Functioning of the central model

The resources of each testbed are integrated and abstracted before being registered to a unified resource metadata management server (Reg). The unified management and utilization of the resources from different testbeds is carried out by the Set.

The functioning of the distributed model is shown in Figure 5-25.

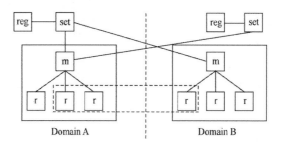

Figure 5-25: Functioning of the distributed model

The resources of each testbed are integrated and abstracted. There is a "borrowing" protocol for the transfer of resources between testbeds.

Currently, these two federated models are being studied in the European Union's FP7. Basic protocols for use in the federated models have emerged, such as SFA (Slice Federation Architecture) and related protocols. At present, both PLANETLAB in the United States and ONELAB in the European Union support SFA architecture, and interconnection is possible.

5.5.5 China's priorities in the construction of a testing environment for future networks

China needs to speed up the construction of its testing environment to bring about technical and service innovations. It must:

- Support dedicated testbeds for core architectural innovation, including the construction of testbeds for innovative network architectures with a certain amount of accumulated experience and a reasonable technical base, such as hierarchical networks, FPBN, and universal networks;
- Integrate existing testbeds and make use of the federated approach in order to establish a sustainable evolving testing environment; utilize the federated approach to integrate existing experimental resources in China to form an experimental environment for the long term; support innovative experimental networks with 20 or so nodes to provide an environment to verify or demonstrate new network protocols or architectures; the Chinese government has an important role to play in the integration of multiple testbeds;
- Make the testing environment interoperable with existing platforms abroad in order to have international influence; consider the possibility of the testing environment becoming the first tier of the next-generation Internet. Connect to FIRE and GENI to provide external access to the testing environment and special projects.

Chapter 6

Technical Solutions for Future Networks

Chapter Highlights:
- *SDN (Software Defined Networking)*
- *NDN (Named Data Networking)*
- *NEBULA*
- *XIA (eXpressive Internet Architecture)*

Overview

The United States began making strategic maneuvers in the future networks field at the beginning of the 21st century. Continuous funding for future networks technology research has been provided, and the National Science Foundation (NSF) launched the GENI program, the FIND initiative, and the FIA project successively. At present, the United States is the world leader in research into future network technologies, and it has a major influence on the future of global network technology.

1) The United States' strategic intent for the future of the Internet field is becoming clearer

Thanks to its first-mover advantage in the Internet field, the United States has been the global leader in information communications since the 1980s. In the course of the evolution of the Internet, the United States has held two priorities. Firstly, it hopes to consolidate and strengthen its dominance in the Internet field by maintaining the existing governance structure and sticking with the existing technology systems. It also recognizes the outstanding issues facing the Internet at present, as well as competition from countries in the EU and the Asia-Pacific region in the development of future networks. As such, it is speeding up its efforts to move ahead with its future network plans, and is enhancing its innovation and experimentation in future network techniques. In doing so, it hopes to maintain its advantage during the long-term evolution of the Internet.

2) The United States has clear strategic priorities in the field of future networks

In recent years, the United States has striven to improve the deployment of its future network strategies. A total of five strategic priorities with corresponding technical directions have been formed.

(1) To create new vitality through technical innovation to promote economic development. The United States has integrated technical network innovation closely with industrial transformation. Through technical innovation, it has opened up basic closed communication networks, so that more status information and control capabilities are made available to Internet applications, thereby boosting innovation. In so doing, new industrial layers and service models will also appear, supporting internet companies to take on more major roles in the development of the Internet economy and industrial transformation. Based on this strategic consideration, the United States supports R&D in techniques such as OpenFlow and Software Defined Network (SDN) for project set-up. By separating the control layer from the forwarding layer of the network, more control of the network can be made available to applications so that they can customize the network according to demand. Ceding some control of networks allows internet companies to make use of the basic networks provided by traditional network operators to build inexpensive but highly efficient networks that can satisfy their needs, promoting the innovative development of services such as cloud computing and the Internet of Things. Meanwhile, it also provides traditional telecommunication companies with a technical avenue to build smart pipelines, and creates opportunities for traditional network operators to transform and develop.

(2) To improve the Internet's application infrastructure and network capabilities through technical innovation in future networks. With the development of data-intensive services such as video-streaming, Internet application infrastructure (with the typical characteristics of content distribution and traffic optimization) has received increasing attention from the US government. Related scientific research projects have become the focus of governmental support, such as the Named Data Network (NDN) and Internet Indirection Infrastructure (I3). Through content routing, these projects implement content-centric networks with capabilities such as computing, storage, and distribution in the network architecture so that a content-centric application infrastructure layer is added between the network layer and the application layer. This would refine the Internet industrial chain, greatly enhance the Internet's content

distribution capabilities, and significantly improve user experience.

(3) To support the development of the Internet in space and outer space through technical innovation of future networks. In competing for control of space and outer space, the United States is speeding up its R&D in space-related Internet technologies. The aim is to address the communication conditions of large delays and high losses between planets, in order to support the interconnection between space and Earth, the connection to satellites and space vehicles, and the development of networks in outer space. The United States is leading the completion of space-related Internet technical standards. It has organized and conducted a technical trial, and has gained a first-mover advantage in the field.

(4) To enhance network security mechanisms and support national security through technical innovation. Taking into consideration the growing conflict between the lack of a trustworthy architecture for Internet security and the need to ensure national security and protect the privacy of individuals, the United States has integrated technical network innovation closely with national security so that security mechanisms embedded in the network are used to enhance the security of the basic network. In this way, the United States is attempting to ensure network security via the network architecture by changing the current security mechanism, which relies on plug-ins. Change would be brought about through digital signatures between communication entities and adding a new network-security sublayer in the network architecture to implement embedded security mechanisms. Meanwhile, the US has also invested in the implementation of the Shadow Network program, in which self-organizing wireless network techniques are used to open up new fronts for overseas penetration. An example is the "Internet in luggage" and "shadow mobile communication network" concepts, in which "luggage" is used as a control center to arrange mobile phones and other terminals into a wireless network for surveillance and connecting organizations.

(5) To attach great importance to the testing of future-network techniques, thus acquiring the first-mover advantage in control of future networks. Considering that technical innovation is currently very active in this field, to avoid the risk of "technique gambling", the United States is hedging its bets by constructing future-network testbeds to cultivate a fertile environment for technical innovation. Future-network testbeds make use of a unified network operating system, and existing network resources are integrated in a federated manner to provide an environment for the testing of various technical network solutions in parallel. Thereby, coexistence and competition among the various network techniques in the testbeds are encouraged. The winning technical solution will make use of the testbed as network core layer (Tier 1), to become interoperable and integrated with the existing networks. The United

States will then replicate the successful model for evolving the ARPA network into the current Internet, and will incubate a prototype future network, thus seizing control of the field. The United States has already integrated existing experimental networks such as Internet2 and PlantLab through the GENI project, and has built a future-network testbed consisting of more than 50 networks and thousands of interconnected network nodes worldwide. The testbed has attracted numerous network technical solutions, giving the United States the first-mover advantage in strategic maneuvering.

3) Future-network technical innovation is proceeding rapidly, and some outcomes are already being used.

The European Union and countries in the Asia Pacific region have also set their sights on the historical opportunity brought about by the transformations in Internet technologies. They have enhanced their technical innovation and testing capabilities in order to contend for dominance of cyberspace infrastructure. Looking at the global development of future-network technologies, coming up with an internationally-recognized architecture that is technically comprehensive is a long and arduous task that may take up to two decades. However, a pool of new techniques can be formed from the outcomes of future-network research. Some of these techniques can be combined with existing networks to have a significant impact on the evolution of existing networks in a relatively short period of time. At present, the separation of control and forwarding in networks, the separation of labels and addresses, and the separation of sending and receiving have become important technical directions and ideas in future-network research. The United States supports R&D in the OpenFlow technique within the GENI project, and this has given rise to the SDN technique, which has been industrialized. In 2012, Google applied it on a large scale in its internal backbone networks. Tencent and China Mobile also started corresponding networking experiments. It is expected that within the next two to three years, the large-scale application of SDN equipment in existing networks will be considerable.

6.1 SDN (Software Defined Networking)

Currently, SDN is the hottest topic in the industry. Academics and practitioners alike are paying close attention to its possible impact. However, experts from different research fields interpret it differently due to variations in their areas of concern and knowledge structures. This section attempts to predict the impact of technical development of SDN by analyzing its essence, characteristics, application areas, and development trends.

6.1.1 Different opinions about SDN

Within online forums, technical seminars, and standards seminars, experts who are focusing on SDN can be roughly divided into three categories: personnel in IDC design, operations, and maintenance; personnel in data equipment design and R&D; and personnel in future-network research and experimentation. The starting points of these three groups are not the same, and their understandings of SDN are also different. As such, their visions and expectations for development are also different. While they all appear to be discussing SDN, the underlying differences in their understanding tend to be quite significant, causing them to operate on different wavelengths.

After the emergence of a new technology, some experts are keen to determine who was behind it, as well as when it was put forward. Usually, they follow up by clarifying that it is actually not an original concept. Even though such discussions have academic significance, they do not help substantively in understanding the essence of the new technology. In discussing the concept of SDN, it is best to start with the market's actual needs for it.

1) The commercial need for SDN first appeared within data centers

To support the migration of virtual machines on the application server, the internal IDC network is usually a Layer 2 network instead of a Layer 3 network. If a Layer 3 network is used, the IP address of the application service to which the virtual machine corresponds also has to change, causing difficulties in service deployment and management. A Layer 2 network does not cause such issues.

However, direct application of the existing Layer 2 networking technique within the IDC causes two problems. First, in Layer 2 networks, a Spanning Tree Protocol (STP) is usually used in order to eliminate loops in the broadcast packet. A logical tree is established between network nodes, and traffic between nodes is transmitted using this tree-shaped topology. Even if there are multiple physical links between network nodes, only one link actually transmits data. The other links are idle, serving only as backups. However, in the IDC, there are frequent data-exchange requests among multiple servers. The STP-based tree-shaped network topology cannot efficiently support such horizontal traffic flow, and the idle links between servers are also a huge waste of network resources. Therefore, STP-based Layer 2 networks are too simple for the IDC, and this needs to be changed. In particular, with the development of cloud computing, the internal Layer 2 networking requirements of IDCs become increasingly pressing.

Second, the IDC usually has many application servers, some of which number tens of thousands or even hundreds of thousands of units. Layer 2 switches need to use ARP and other protocols to learn the source address of received data packets to establish a

MAC address table. Due to the large number of application servers, there are many MAC address table entries, which usually exceed the capacity of the conventional MAC address table of the Layer 2 switch. This results in a large number of MAC addresses failing to enter the MAC address table. If the data frame corresponding to a MAC address cannot be found in the MAC address table, the Layer 2 switch will broadcast in the Layer 2 domain. This causes a flood of traffic within the Layer 2 network, affecting the efficiency of the internal IDC network. The root cause of the two problems mentioned above is the overly simple design of the conventional Layer 2 network. The Layer 2 switch only learns the MAC address; it does not plan the data forwarding path based on it. In other words, there is no control plane in conventional Layer 2 networks (or rather, the control plane is very weak and merges with the forwarding function), and only the data plane (responsible for forwarding data frames) is present. Therefore, adding a control plane (or enhancing the control plane functions) in a Layer 2 network that is responsible for traffic scheduling and management between the internal nodes of large Layer 2 networks becomes an urgent need. The mainstream solution at present is to use variations of the IS-IS routing protocol to build the control-plane routing function. The OpenFlow technique is then used to define the interface between the control plane and the forwarding plane. This gives rise to the idea of separating the control plane from the forwarding plane. However, this is only valid for Layer 2 networks.

2) The commercial demand for SDN comes from optimizing the internal functions of routers

In conventional routers, the interface between the control plane responsible for route-planning and routing policies, and the data plane responsible for data encapsulation and high-speed forwarding, is closed and tightly coupled. Each manufacturer uses its own protocol or interface to connect to both the control plane and the forwarding plane. This is also why leading manufacturers such as CISCO and JUNIPER are able to maintain a technical barrier and to push emerging competitors aside.

However, there are two forces that are quietly challenging this model. One is the large Internet companies that need to build their own corporate networks and believe that the current communication requirements of their own corporate networks are very specific. However, the functions of conventional routers are too complex, with more than 80% of their functions and features not used in the abovementioned corporate networks. As such, these companies believe that it does not make any sense to pay for such useless functions, and there is a demand for independently-designed routers that are simple and efficient. This is why companies such as Facebook, Google, and Yahoo initiated the ONF (Open Networking Foundation) to develop SDN standards. As these Internet companies have had a great deal of success using application servers they customized within IDCs, they are confident in their ability to develop efficient routers on their own. The other force is the emerging data equipment manufacturers who are attempting to form an open and standardized device interface by decoupling the control plane and the data plane in the router from each other, so that control functions can be centralized and removed independently. By doing so, data forwarding equipment can be made more generic and simple, and costs could be lowered. This can help to break the oligopoly held by CISCO, JUNIPER, and other manufacturers, leading to new opportunities for development.

Based on this consideration, the IETF has conducted research into separating the control plane and the forwarding plane in the router. The FORCES task force was established to define the communication protocol between the control plane and the forwarding plane in the router. Although the concept proposed this time is the same concept of separating the control plane from the forwarding plane, it involves the separation of the control plane and the forwarding plane in the Layer 3 network.

3) The commercial demand for SDN comes from future-network research and testing

In order to solve the current problems of insufficient network address space, poor service quality, a lack of security mechanisms, and poor network manageability and controllability faced by IP networks, future-network researchers are looking into new network architectures and key technologies. Although there are many research directions, there is no clear consensus for any technical pathway in particular. However, even though the technical pathway is not clear and there is an endless stream of new solutions, it is necessary to establish an ultra-large testing environment for new future-network technologies. This experimental network will have the flexibility of providing various technical solutions with a resource-independent test environment that is used for incubating the preferred ones. The United States and Europe established the GENI and FIRE networks respectively for this purpose. In the construction of the experimental network, the designers hope to be able to flexibly control and deploy routing protocols on the network nodes to achieve efficient forwarding. As a result, there is an increasingly strong need to separate the control plane from the forwarding plane for the nodes in the experimental network. The separation of the control plane will gather network control functions together, making them smarter and rendering the protocols of network forwarding-functions irrelevant while the functions themselves become highly efficient.

Separating the control plane from the forwarding plane for the nodes in the experimental network means that software can be used to design and configure the nodes involved whenever a new type of network architecture and solution appears. This allows the new network form to be implemented quickly, thus supporting network technical innovation and verification efficiently.

Other than the three SDN demands and their corresponding experts, there are also some experts who subscribe to the view of SDN as a form of unified and intelligent network management. They are thus committed to realizing a unified network-control system that is able to intelligently manage multiple network devices. For example, for LTE deployment, in designing the IP RAN, an integrated network-management system can be used to configure and manage multiple simplified routers at the periphery in order to improve the efficiency of network-policy deployment. However, this understanding separates the management plane from the control plane and the data plane, and is not the separation of the control plane and the forwarding plane. As such, it should not be understood as SDN technology.

6.1.2 Understanding the essence of SDN

The three developmental needs of SDN were analyzed in the previous sections. On the whole, it can be said that they stem from a pursuit of network openness.

The opening up of networks is an inevitable trend in the industry's development. Not only can it bring about more efficient equipment and networks, but the industrial chain can also be segmented further, bringing new opportunities for industrial development. There is a precedent for openness in mechanical parts. The standard interchange between parts has refined the industrial chain for machinery, improved the production efficiency of completed machinery, and propelled the industrial revolution forward. In the field of network communications, similar expectations can be achieved through SDN technology.

In summary, there are two planes within a data device, as shown in Figure 6-1.

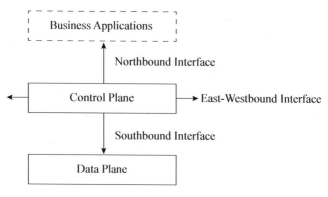

Figure 6-1: Separation of two planes

To understand SDN from the perspective of network openness, SDN can be divided into three categories. The openness between each category is an increasing relationship.

1) SDN that opens the northbound interface of the control layer (also known as SDN-N)

This idea investigates the openness of the northbound interface between the control surface and the business applications of the data network, providing upward resource abstraction, and implementing a software programmable and controllable network architecture. The openness of the northbound interface of the control layer in the data network is conducive to the Internet application service sensing the data network state, optimizing the application design of the service, and improving the user experience. Therefore, it is supported by the Internet service provider. The northbound interface openness study originated during the P2P research boom five years ago. In order to achieve coordination between P2P traffic optimization and data network traffic scheduling. The IETF initiated several task forces such as ALTO and DECADE. As the P2P heat has subsided, the pace of this research has slowed, but the escalation of SDN has injected new vitality into this corner of the field. Investigations into the openness of northbound interfaces mainly comprise abstracting the common characteristics of different business applications and their needs for the bearing of data networks. However, the diversity of business applications means that the current progress of this work is not smooth. From a purely academic point of view, the opening of the northbound interface does not involve the separation of the control layer and the forwarding layer. Therefore, some experts believe that it is a technical idea and thus not part of SDN-related research.

2) SDN that opens the southbound interface of the control layer (also known as SDN-S)

This idea is usually understood as being a separation of the control plane and the data plane in the data network. The current ONF OpenFlow protocol and the IETF Forces protocol both work at this level. They define the communication protocol between the control plane and the data plane after separation. The difference between the OpenFlow and Forces protocols is that the forwarding device hardware faced by OpenFlow assumes that only ten tuples are supported. OpenFlow can configure various forwarding rules for ten tuples. Forces assumes that the forwarding device hardware it faces is protocol-independent. Forces can arbitrarily define the processing logic of the underlying forwarding device in the XML language format. Protocol-independent forwarding devices have also become a research hotspot. To be protocol independent requires hardware to be equipped with many functions. This seems to be a very difficult task, but some chip and device manufacturers have developed protocol-independent forwarding products. It is a direction worthy of attention.

3) Opening up the interfaces along the east-west axis in the control plane of the SDN (also known as SDN-SE)

After opening up the interfaces along the north-south axis, another problem faced in SDN development is the scalability of the control plane, i.e. how the control planes of multiple devices collaborate. This involves defining the interfaces along the east-west axis in the control plane of SDN. Without doing so, SDN is at most an optimization technique within the data device, and different SDN devices would still depend on the IP routing protocol for interconnection. This would severely limit the impact of innovations in network architecture. Defining standards for the interfaces along the east-west axis in the control plane will allow SDN devices to form a network, so that the SDN technique can spread beyond the confines of IDCs and data devices to become a network architecture with a profound impact. At present, research into this aspect of SDN has just started, and both the IETF and ITU have not stepped into the field.

Judging by the development trend for network openness, the impact of the SDN concept on the design of network equipment and network architecture is still in its infancy. The subsequent standardization of interfaces along the north-south and east-west axes in the control plane of SDN, as well as the integration of SDN and network virtualization techniques, enable the SDN technique to exhibit greater vitality and have a further reaching influence.

6.1.3　A long road ahead

As mentioned earlier, if the SDN technique and concept were to remain within the internal networks and data devices of IDCs, its impact would be limited. It would only affect traffic optimization, data device structure, and performance optimization within the Layer 2 network. However, if the northbound interfaces (especially the east-westbound interfaces in SDN) are opened up and standardized, it would have a profound impact on the structure and business model of the communications network. The impact of SDN that currently has the industry buzzing is based on this very premise. The impact of SDN based on SDN-SE is discussed in the following paragraphs.

(1) Internet companies. On the one hand, the separation of the control plane and the forwarding plane provides an opportunity for the transfer of network control rights. Currently, Internet giants such as Google are bypassing network operators to

build their own network infrastructure by engaging in activities such as purchasing submarine communications cables, deploying fiber-optic networks, and setting up "hot-air balloon" networks. The advent of the SDN technique offers Internet service providers (ISPs) the technical means and opportunities to build inexpensive and efficient networks. This is an ongoing challenge to conventional network operators, and may have a significant impact on the development of the Internet's business model. Meanwhile, the SDN technique optimizes the internal networks of IDCs, which contributes significantly to improving their efficiency, and reducing the costs of construction, operation, and maintenance. Companies such as Google have successively released examples and data that show significant improvements in IDC efficiency after using the SDN technique. Network optimization in IDCs has a direct effect on promoting the development of cloud computing.

(2) Equipment manufacturers. For emerging manufacturers, SDN means new market and business opportunities within the "blue ocean strategy". This is why non-mainstream data equipment manufacturers such as NEC are rushing to develop SDN routers and switches before the traditional big players such as CISCO. Meanwhile, for traditional manufacturers, SDN means a breaking up of the monopoly, and is of great significance for stimulating technical innovation and competition and accelerating the development of network technologies.

(3) Network operators. SDN provides the technical means to build smart pipelines. Network operators can use it to implement network optimization and efficient management, improve the network's intelligence, controllability, and manageability, and significantly reduce the costs of network construction, operation, and maintenance. Meanwhile, SDN can facilitate the opening up of the underlying networks by network operators and promote the optimization and innovation of Internet service applications. For example, with the support of the OpenFlow network, IaaS users can set their own paths and security policies for data flow within the local network, instead of having control over several virtual devices.

(4) Research and exploration into future networks. Driven and influenced by SDN and other techniques, network virtualization and the separation of control and forwarding have become the basic characteristics of the network architectures of the future, thus promoting innovation in the field. Meanwhile, SDN provides the basic technology for constructing large-scale testbeds for future networks. In experimental networks, SDN can provide new technical network solutions with relatively independent resources, as well as network control functions. That is to say, the routing paths and forwarding switching-rules for data in experimental networks can be freely defined, which is beneficial for technical innovation.

In addition, some experts have suggested that the SDN technique be used in network security for the deployment of security control and management nodes in base stations (or POPs) of fixed networks. This would connect the controller of the security control and management nodes and establish network security boundaries (a UNI/NNI interface). Doing so has several advantages. Firstly, admission control can be implemented to shield the network from user attacks, as the core IP network is unreachable. Secondly, it facilitates the standardizing, issuing, and deploying of security rules to implement group defense and group management. Thirdly, it can be combined with cloud computing to discover abnormal traffic autonomously, and to comprehensively analyze the characteristics of this traffic, thereby forming a spontaneous immune system for the network. In this way, detection by a node could lead to immunity for the entire network. Fourthly, the interconnection of these nodes forms an emergency communication system for the network.

6.2 NDN (Named Data Networking)

A content-oriented architecture—the Named Data Network (NDN)[14][15] is one of the five Future Internet Architecture (FIA) projects funded by the National Science Foundation (NSF) in the United States. Currently, the Internet's main service is content distribution. Data access is concentrated, and as conventional Internet architecture is designed with end-to-end transmission in mind, data heat-sensing and intelligent scheduling are lacking. The mismatch between the user service model and the end-to-end traffic scheduling of the network results in a large amount of duplicated data transmission in networks. This leads to multiple copies of the same content and repeated transmission of the same traffic, further exacerbating the burden of network traffic. To solve this problem, NDN proposes a new network architecture design that alters the basis of network communication from the current network address or location to network content. This would solve existing issues with Internet traffic, such as scalability, security, and dynamism.

6.2.1 NDN Architecture

NDN also makes use of the hierarchical, hourglass structure of TCP/IP networks. The only difference is that the "IP packet layer" at the narrowest point in the TCP/IP architecture becomes the "named content chunk layer", as shown in Figure 6-2. In addition, two new layers are introduced in NDN, namely the "strategy layer" right below the narrowest point, and the "security layer" right above it. The "strategy layer" makes forwarding and caching decisions to implement mobility, multi-path transmission, and network caching, to better utilize the multiple concurrent connections underneath it (such as the Ethernet, 3G, and 802.11). The "security layer" implements content-based security. This is unlike the existing TCP/IP networks that only implement security protection for the transmission channel. This will help avoid the various host-based network attacks the existing Internet faces.

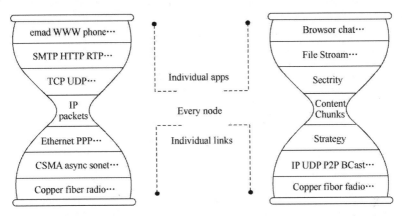

Figure 6-2: NDN adopts the hierarchical, hourglass model used by TCP/IP (the "IP packet layer" at the narrowest point becomes the "named content chunk layer")

6.2.2 NDN Node Model

In the NDN, each data block has a unique name. The NDN uses a "Push-Pull" communication model, which is driven by both the content generator and the consumer. There are two types of packets in the network: the interest packet and the data packet (see Figure 6-3). When consumers need to obtain certain data, they broadcast an interest packet through all their available network connections to request it. If a network node that receives the interest packet has the data requested, it responds to the request through the data packet. The interest and data packets have a one-to-one correspondence with the names of the data. The data packet is only transmitted in response to the interest packet it corresponds to. It will also use up the interest packet.

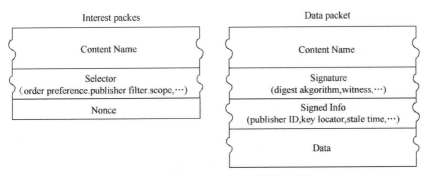

Figure 6-3: Two types of NDN packet

An interest packet that has already received a response from a data packet is no longer transmitted in the network.

The basic operation of an NDN node is similar to that of an IP node. After a packet reaches an interface of a node, a longest-match lookup is performed using the name of the packet before the action corresponding to the search outcome is performed. Figure 6-4 shows the model of an NDN packet-forwarding engine, which includes the three main data structures of the FIB (Forwarding Information Base), CS (Content Store), and PIT (Pending Interest Table).

The FIB table is used to forward Internet packets to potential data-source nodes with matching data. Other than the possibility of more than one output interface for the FIB table in NDN, the functions of the FIB table in NDN are largely similar to the FIB table in IP. This is reflected in NDN in that the forwarding path of Interest packets does not have to be a spanning tree. Interest packets can be forwarded to multiple data-source nodes, and data can be requested from these nodes in parallel.

The CS is used to cache data, thus adding cache function to the network. The CS in NDN is similar to the buffer cache in IP routers, only that a different replacement strategy is used. Since each IP packet belongs to a point-to-point conversation, the packet will be meaningless after being forwarded. As such, forwarded packets are forgotten by the IP, and their buffer is immediately reclaimed. In NDN, the packets are idempotent, self-identifying, and self-verifying, so the content of each packet may be reused repeatedly by other users (e.g. many users reading the same newspapers or watching the same YouTube video). In order to maximize the probability of sharing, and also to minimize the bandwidth and delay of downstream transmission, NDN nodes will keep the received data as long as possible (such as using the LRU or LFU replacement algorithm).

The PIT is used to record the uplink path that the interest packet has forwarded to the data-source node, so that the requested data packet can be sent to the requester in the reverse direction. The PIT of each NDN node records the source ports that have been forwarded, but the corresponding interest packets have not been received. In NDN, only the interest packet is routed. When an interest packet is forwarded to a potential data source, a trail is left behind. A data packet will follow the trail in the opposite direction to make its way to the data requester. Every PIT is a trail. As for an interest message for which no matching data packet is found, its corresponding PIT will eventually time out. If the sender of the request still wants the data, it will resend the interest packet.

When an interest packet reaches a port, the router will perform a longest-match based on its ContentName. The order of the search is: CS table, PIT table, and FIB table. If a matching entry is found in the CS table, the data packet is sent directly to

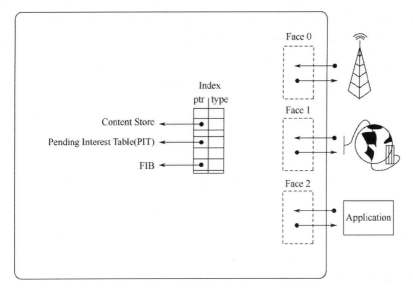

Figure 6-4: The NDN packet-forwarding engine model

the port that received the interest packet. The interest packet is then ignored, since it has already been responded to.

Otherwise, if there is an exact match in the PIT table, the port receiving the interest packet is added to the forwarding list of the corresponding entry of the PIT table, and the interest packet is then ignored. In the current system, one or more interest packets requesting the same data have been forwarded, and await the transmission of the corresponding data packet. When the data packet is transmitted to the router, the router is able to forward it to all the ports in the forwarding list.

If there is no exact match in the PIT table but a match in the FIB table, the interest packet needs to be sent to the data source according to the sending list of the FIB table entry to request the data packet. Note that if the port reached by the interest packet is also in the matching entry of the FIB table, the port should be deleted first before the interest packet is forwarded to the remaining ports.

If there is no match in the FIB table, the interest packet is discarded. This node does not have any matching data and does not know how to find the data required.

The processing of data packets is relatively simple because they are not routed. Instead, they are forwarded to the original requester in the opposite direction to the order in which the ports are recorded in the PIT entry. When the data packet arrives, a longest-match is also carried out. If there is matching content in the CS table, it means that the data packet is a duplicate and is discarded. If no match is found in the PIT table, the packet has not been requested and is also discarded. If there is a match in the PIT table, it means that the data packet was requested. The data packet can then be verified (optional) and added to the CS table. The requesting-port list of the matching entry in the PIT table minus the set that is to receive data packets gives the forwarding ports of the data packet.

This kind of data acquisition method based on interest packets has multipoint characteristics, which provides flexibility for maintaining communication in a highly dynamic environment. Any node that accesses multiple networks at the same time can serve as a content router for the multiple networks. By using its cache, the mobile node can become the transmission medium for two unconnected areas, or provide a delayed connection (similar to DTN) for timed-out links. This interest/data packet exchange mode only comes into play when there is only a local connection. For example, two colleagues working on their notebooks can share documents over a local ad-hoc network even when they are not connected to the Internet or a corporate network.

6.2.3 Features of the NDN Technique

In NDN, network architecture is redesigned to incorporate user requirements. In this architecture, the narrowest point of the funnel model has been defined as the Named Content Chunks. At present, Internet applications must rely on a complex middle layer to translate the abstract content of the IP-specified host into the required content. NDN greatly simplifies the process of developing application programs. In contrast, new applications will also be further developed.

NDN-signed data provides the foundation needed for credibility in future networks. Applications can create fine-particled and self-defined models for authentication, authorization, and credibility.

In an NDN data packet, the signature is used to verify both the integrity and the source of the data. After a user has verified the signature in a received data packet, they are assured that what they have received is a copy of the original data from the correct publisher. Therefore, NDN achieves secure data transmission without requiring that publishers and clients communicate directly with each other.

Through matching the data packet to the interest packet on each path, the NDN provides a strong network-traffic balancing solution that handles every hop. Therefore, NDN networks can manage unicast and multicast traffic on their own without relying on transport protocols. In addition, NDN separates routing rules from the forwarding mechanism.

NDN promotes competition and the freedom to choose in networks by empowering users. Some network economy models have shown that monitoring the performance of information delivery is key to ascertaining the responsibilities of the ISPs. However, in today's global routing system, IP only chooses a single path to reach the destination node. This path is usually asymmetric due to "hot-potato" routing. This means using the services provided by several service providers to reach the destination address, thus making it difficult to measure and compare the performances of different service providers. In contrast, the multi-route forwarding capabilities and feedback loop built into NDN allow users to explore multiple paths and monitor delivery performance before making the final decision. For example, when there are multiple users and small service providers at the network connection interface, the user is free to choose the best-performing service provider. This will encourage innovation and investment in the network architecture through competition.

NDN democratizes the distribution of content. This is another way in which it significantly promotes choice and competition. The dissemination of content and knowledge is an important social function of the Internet. Although today's Internet is undoubtedly a success, its ability to disseminate information is clearly insufficient to deal with the amount of content generated every day. The caching capabilities built into NDN allow information generators (whether content providers or individual users) to efficiently disseminate information to a global audience without requiring any special architecture (such as CDN). This will have far-reaching consequences, especially for underdeveloped regions and those who wish to freely and fully express their views. This is also a positive element, as it encourages people to create and produce original content.

6.3 NEBULA

NEBULA[17][18] is another of the five major FIA projects funded by the NSF. NEBULA ("cloud" in Greek) is an innovation project for new network architecture that connects cloud computing data centers. The NEBULA network architecture is made up of three parts:

- NDP: NEBULA Data Plane;
- NVENT: NEBULA Virtual and Extensible Networking Techniques;
- NCore: NEBULA Core.

1) NEBULA Data Plane (NDP)

The NDP is a network protocol. Each packet in this protocol contains the following four parts for each management domain it passes, as shown in Figure 6-5.

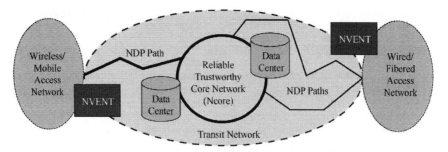

Figure 6-5: Model of the NEBULA architecture

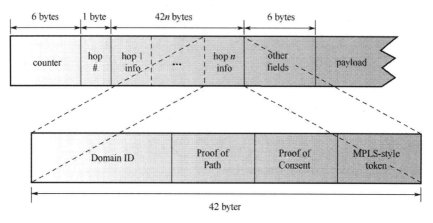

Figure 6-6: NDP packet format

The domain identifiers are as follows:

- PoC, which proves that the management domain has authorized the path;
- PoP, which proves that the packet follows the path it illustrates;
- An identifier for multi-protocol label switching.

Research in the previous phase found[19][20] that these elements are not only sufficient for the network elements to express their own rules, but also to execute the corresponding rules. When a packet reaches an autonomous domain, this domain has enough information to decide whether to deliver the packet. Checking the PoC will determine if the packet is authorized; checking the identifier will determine what internal resources the packet will consume, and which intermediate components the packet has passed through; checking the PoP will determine whether the packet is disseminated on the authorized path.

Preliminary experiments and prototypes have proved that although this architecture takes up space in the data packet and increases the amount of processing by the data platform, it is highly flexible.

2) NEBULA Virtual and Extensible Networking Techniques (NVENT)

NVENT's task is to collect available resources and return the available paths and other related information when an application or provider requests a service that has certain requirements.

More specifically, NVENT's job is to determine the values of various parameters in the NDP packet. In particular, NVENT is responsible for deciding the propagation path of the packet, obtaining permission from all intermediate components, and understanding the identifiers required on this path. Normally, the sender can request this information from the NVENT server before placing it in the packet. However, it is more common for the sender to send the same information as the Internet at this point in time. The proxy server or gateway then requests information from the NVENT servicer and converts these packets into NDP packets. In this way, when an NDP packet enters the network, each autonomous domain has enough information to check it, as shown in Figure 6-7.

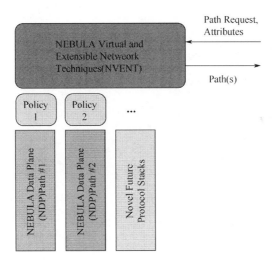

Figure 6-7: NVENT service interface for path selection

3) NEBULA Core (NCore)

NEBULA's core networks will be built on the next generation of core routers that will provide the fastest transmission speeds and will be readily available. Widespread availability requires that a next-generation router's control plane software be a fault-tolerant distributed system[21]. Since a single-core CPU cannot handle the data-forwarding in the first-layer ISP, next-generation high-performance routers will exist in a distributed form. In future, routers will support multiple slots, each with multiple network cards, forwarding servers, and control servers. The components of the router will be connected by high-speed optical fibers. From the outside, the entire router is a small distributed system, as shown in Figure 6-8.

In addition to high-speed and stable routers, the core network requires high-speed connections between data centers and routers. The NEBULA team has begun working with CISCO and Intel to explore parallel connections between data centers and routers[22] to provide reliable high-speed connectivity. This will solve the problem of speed mismatch between the high-speed connections of the grid storage and computing in the data centers, and the WAN in the core network. At the same

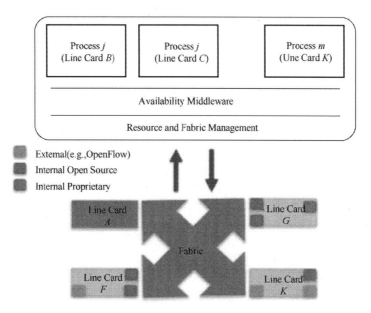

Figure 6-8: Future hardware and software architecture of core routers

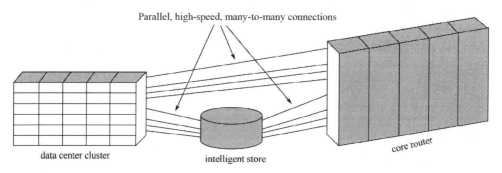

Figure 6-9: Parallel multiple-to-multiple connections between data centers and the core network

time, the use of multiple paths[23] between the source and destination addresses can greatly improve the transmission bandwidth, and also allow for better handling of the occurrence of line or router failures in the network. In NEBULA, a multi-path routing protocol compatible with the NDP mechanism will be developed, as shown in Figure 6-9.

6.4 XIA (eXpressive Internet Architecture)

XIA (eXpressive Internet Architecture)[24][25] is another of the five FIA projects funded by the NSF in the United States. Its aim is to address the growing diversity of Internet application models and trustworthy communication issues. In so doing, it attempts to create an independent network that can internally support communication between different communication entities (such as hosts, content, services, and the various new communication entities that may appear in future). XIA defines a specific "narrow point" for each type of communication entity, establishing its API for communication, network communication mechanisms, and endogenous security mechanisms.

6.4.1 Technical Ideas for XIA

The academic world has proposed various types of communication methods, such as content-centric[14], service-centric[26][27], multicasting[28], and mobility support[16]. XIA researchers propose a framework that can support some or all of the abovementioned functions. It should allow a choice to support these functions or not, both anytime and anywhere.

To this end, researchers have proposed the XIA network architecture. It adopts the hourglass structure of the existing Internet, but it differs in many aspects. Firstly, the current Internet is host-centric, but XIA supports multiple network entities, including hosts, services, and content, as well as entities that may appear in the future. Next, the XIA supports the evolution of network architecture, and provides a mechanism for deploying new functions incrementally. XIA is based on three core concepts:

1) Multiple network entities

In the host-centric TCP/IP architecture, there is only one type of network entity: the host. However, XIA supports a variety of network entities, including hosts, content, and services, as well as entities that may appear in the future. It also supports the incremental deployment of new entity types. An application can use several types of entities to convey its needs to the network, instructing routers to perform specific processing on the data packets. In addition to a more flexible interaction mode, network optimization is also easier to implement, as the underlying functions of the network are used directly by applications.

2) Flexible address resolution

XIA supports the evolution of network architecture and the mechanism for the incremental deployment of new entity types. It introduces a new address structure that allows legacy systems to be compatible with new entities through flexible address resolution. All network entities in the XIA architecture have a corresponding ID. When an application requests the functions, content, and services provided by the network, the entity ID is specified in the address field of the data packet. In the TCP/IP architecture, there is only one source address and destination address each for a data packet, i.e. the host's ID (IP address). For the address field of an XIA data packet, there can be more than one entity ID. One is the primary entity, while the rest are backup entities.

When a router receives a data packet, it parses the packet's destination address in the following way: First, it looks for the primary entity. If the router supports this type of primary entity, it processes the data packet directly according to the protocols relevant to the entity type; otherwise, it will proceed to check the backup entities. There can be more than one backup entity, and the parsing method is the same as mentioned above, i.e. if the current backup entity is not supported, then the next one is checked. When a new type of entity is introduced into the network architecture, it is regarded as the primary entity in the address field of the data packet. At the same time, an entity type that has been widely supported is then listed as the backup, so that even if the router is unable to identify the new entity type, it is able to process the data packet correctly based on the backup entity.

3) ID with built-in security

XIA requires that all entity IDs in the network have built-in security. The method of generating the entity ID is related to the entity type. The entity ID adopts the hash value of the host's public key; the security of the content obtained is achieved through verifying the accuracy and integrity of the content, with the entity ID adopting the content's hash value. Through IDs with built-in security, network entities (i.e. hosts and routers) are able to verify the security of the entity directly without having to access external databases or configuration information. IDs with built-in security provide a basic security mechanism on which higher-level security mechanisms can be developed.

6.4.2 XIA entity type

In the XIA architecture, the network can simultaneously support many different communication methods by using multiple entity types. Thereby, the expressiveness of the network architecture is enriched. XIA adds "entity type" to the network architecture, which must include the following three aspects:

(1) The communication semantics of the entity
The semantics of the entity type are defined, i.e. the purpose and meaning of communication using the said entity type. The most common purposes are to obtain specific content and to communicate with a specific host. When a communication method cannot be effectively expressed with an existing entity type, the defining of a new entity type can be considered.

(2) The entity ID, method of generating said ID, and method of verifying its security
The method of generating the entity ID and its security rules are related to the entity's semantics. For example, content retrieval makes use of the content hash-value, and communication with a host uses the hash-value of the host's public key. Based only on the different types of entities, XIA has identifiers such as the host identifier (HID), the service ID (SID), and the content identifier (CID).

(3) The processing of the entity's data packet by the router
It must be ensured that intermediate devices in the network (such as routers) can process data packets correctly. In accordance with the semantics, the router can optimize the processing of data packets (such as cache content) to support anycast.

References

[1] Hierarchical Switch Network Architecture, Hualin Qian, Jingguo Ge, Jun Li, Tsinghua University Press, Beijing.

[2] Basic Research on Integrated Trusted Networks and Universal Service Systems: Goals, Ideas and Progress, Hongke Zhang, Hongbin Luo. 2008.12.

[3] Research on Ontology Merging for Semantic Web Services, Zhenghai Luo, Master's Thesis.

[4] Research on Key Technologies and Methods of IPv6 Network Address Planning and Real-name Trusted Communication, Tao Chen, PhD Thesis, 2010.

[5] ITU-T Y.2601, Fundamental characteristics and requirements of future packet-based networks.

[6] ITU-T Y.2611, High-level architecture of future packet-based networks.

[7] ITU-T Y.2612, Generic requirements and frameworks for addressing, routing, and forwarding in future packet-based networks.

[8] ITU-T Y.2613, General technical architecture for public packet telecommunication data networks.

[9] Towards the Future Internet, IOS Press BV.

[10] Named Data Networking (NDN) Project, Lixia Zhang, Deborah Estrin, and Jeffrey Burke, University of California.

[11] Internet Indirection Infrastructure, Ion Stoica, Daniel Adkins, Shelley Zhuang University of California, Berkeley.

[12] The Future of the Internet, Pew Research Center.

[13] An Introduction to Virtualization on Planet Lab, Baris Metin.

[14] Implementing Network Virtualization for a Future Internet, Panagiotis Papadimitriou, Olaf Maennel, 20th ITC Specialist Seminar 2009.

[15] Architectural Trends in Future Communication Systems, Donal O'Mahony.

[16] Future Internet Research and Experimentation, European Commission.

[17] http://ec.europa.eu/research/fp7/index_en.cfm.

[18] http://www.nets-find.net/.

[19] http://www.planet-lab.org/.

[20] Future Internet Architecture Research, China Communications Standards Association (CCSA) Research Topic (2012B81), Institute of Computing Technology, Chinese Academy of Sciences

[21] A Research Summary of Future Internet Architecture, Gaogang Xie, Yujun Zhang, Zhenyu Li, et al. Chinese Journal of Computers 2012, 35 (6): 1109-1119.

[22] http://www.huawei.com/broadband/iptime_backbone_solution/era/100g_transport_era.do, 2011.

[23] The CIDR Report, http://www.cidr-report.org.

[24] Report from the IAB workshop on routing and addressing. D. Meyer, L. Zhang, K. Fall. RFC 4984, 2007.

[25] G. Pallis and A. Vakali. Insight and Perspectives for Content Delivery Networks, Comm. of the ACM, 49 (1): 101-106, 2006.

[26] M. Meeker, S. Devitt, L. Wu. Internet trends. CM Summit, New York, 2010. http://www.morganstanley.com/institutional/techresearch.

[27] W. Gao, G. Cao. Fine-grained mobility characterization: steady and transient state behaviors. ACM Mobi Hoc2010.

[28] N. Azimi, H. Gupta, X. Hou, J. Gao. Data preservation under spatial failures in sensor networks, ACM Mobi Hoc2010.

[29] H. Xie, R. Yang, et al. P4P: provider portal for applications. ACM SIGCOMM2008.

[30] T. Karagiannis, K. Papagiannaki, M. Faloutsos. BLINC: multilevel traffic classification in the dark. ACM SIGCOMM2005.

[31] NSF Ne TS FIND Initiative, http://www.nets-find.net/.

[32] NSF Future Internet Architecture Project, http://www.nets-fia.net/.

[33] FIRE: Future Internet Research and Experimentation, http://cordis.europa.eu/fp7/ict/fire/.

[34] L. Zhang, D. Estrin, et al. Named data networking (NDN) project. PARC Technical Report NDN-0001, 2010. http://www.named-data.net/index.html.

[35] V. Jacobson, D. K. Smetters, et al. Networking named content. Communications of the ACM, 55 (1): 117-124, 2012.

[36] Seskar, Ivan, et al. "The MobilityFirst future internet architecture project." Proceedings of the 7th Asian Internet Engineering Conference. ACM, 2011.

[37] NEBULA: Future Internet Architecture, http://nebula-fia.org/.

[38] Anderson, Tom, et al. "The NEBULA Future Internet Architecture." The Future Internet. Springer Berlin Heidelberg, 2013. 16-26.

[39] Naous, Jad, et al. "Defining and enforcing transit policies in a future Internet." University of Texas at Austin, Department of Computer Sciences Technical Report TR-10-07 (2010).

[40] Naous, Jad, et al. "The design and implementation of a policy framework for the future Internet." Submitted to the 2010 USENIX Symposium on Networked Systems Design and Implementation. 2009.

[41] Caesar, Matthew, et al. "Design and implementation of a routing control platform." Proceedings of the 2nd conference on Symposium on Networked Systems Design & Implementation, Volume 2. USENIX Association, 2005.

[42] Traw, C. Brendan S., and Smith, Jonathan M., "Striping within the network subsystem." Network, IEEE 9.4 (1995): 22-32.

[43] Maxemchuk, Nicholas F. "Dispersity routing." Proceedings of ICC. 1975, 75.

[44] eXpressive Internet Architecture Project, http://www.cs.cmu.edu/xia/.

[45] Dongsu Han, Ashok Anand, et al. XIA: Efficient Support for Evolvavle Internetworking. The 9th USENIX Symposium on Networked Systems Design and ImplementationNSDI'12, 2012.

[46] Nordstrom, Erik, et al. "Serval: An end-host stack for service-centric networking." Proc. 9th USENIX NSDI (2012).

[47] Saif, Umar, and Justin Mazzola Paluska. "Service-oriented network sockets." Proceedings of the 1st international conference on Mobile systems, applications and services. ACM, 2003.

[48] Deering, Stephen E. Multicast routing in a datagram internetwork. No. STAN-CS-92-1415. Stanford University CA Department of Computer Science, 1991.

[49] Rouskas, George N., et al. "ChoiceNet: Network innovation through choice." ONDM. 2013.

[50] Anderson, Thomas, et al. "Overcoming the Internet impasse through virtualization." Computer 38.4 (2005): 34-41.

[51] Future Internet Assembly, http://www.future-internet.eu/home/future-internet-assembly.html.

[52] The FP7 4WARD Project, http://www.4ward-project.eu/.

[53] Constructing the Next-generation Internet Based on Real IPv6 Source Address Authentication Architecture, Jianping Wu, Gang Ren, and Xing Li, Science China: Volume E 38.10 (2008): 1583-1593.

[54] CERNET Project, http://www.cernet.net

[55] Current Research Status of Measurable, Controllable, and Manageable IP networks, and Some Important Development Trends, Meng Luoming, Journal of Communications. 29.12 (2008): 96-101.

[56] Xie, Gaogang, et al. "Demo Abstract: Service-Oriented Future Internet Architecture (SOFIA)."

[57] Planet Lab Europe Project, http:// www.planet-lab.eu/.

[58] Magana, E., et al. "The European traffic observatory measurement infrastructure (ETOMIC)" IP Operations and Management, 2004. Proceedings of the Workshop on IEEE, 2004.

[59] Tighe, Warren. "Network for integrating transportation operations systems (NITOS)". Vehicle Navigation and Information Systems Conference, 1995. Proceedings in conjunction with the Pacific Rim Trans Tech Conference. 6th International VNIS. 'A Ride into the Future'. IEEE, 1995.

[60] DIMES Projects, http://www.netdimes.org/new/.

[61] K-GENI Project, http://groups.geni.net/geni/wiki/K-GENI.

[62] Chun, Brent, et al. "Planet Lab: an overlay testbed for broad-coverage services." ACM SIGCOMM Computer Communication Review 33.3 (2003): 3-12.

[63] Global Environment for Network Innovations (GENI) Project, http://www.geni.net/.

[64] Serge Fdida, Timur Friedman, and Thierry Parmentelat. "One Lab: An open federated facility for experimentally-driven future internet research." New Network Architectures. Springer Berlin Heidelberg, 2010. 141-152.

[65] JGN2plus Project, http://www.jgn.nict.go.jp/.

[66] An Overview of the High-Performance Broadband Information Network in China (3TNet), Jiangxing Wu, Communications World. 2002, 8.12: 37-39.